Second Edition

A C T H E O R Y

In Partnership with the **NJATC**

DELMAR
CENGAGE Learning™

Australia • Brazil • Japan • Korea • Mexico • Singapore • Spain • United Kingdom • United States

AC Theory,
Second Edition
NJATC

Vice President, Technology
Professional Business Unit:
Gregory L. Clayton

Product Development Manager:
Ed Francis

Product Manager:
Stephanie Kelly

Editorial Assistant:
Nobina Chakraborti

Marketing Director:
Beth A. Lutz

Executive Marketing Manager:
Taryn Zlatin

Marketing Specialist:
Marissa Maiella

Production Manager:
Andrew Crouth

Content Project Manager:
Christopher Chien

Art Director:
Bethany Casey

Technology Project Manager:
James Ormsbee

Director of Technology:
Paul Morris

Library of Congress Cataloging-in-Publication Data Card Number: 2007013358

ISBN-13: 978-1-4180-7343-5

ISBN-10: 1-4180-7343-1

Delmar Cengage Learning
5 Maxwell Drive
Clifton Park, NY 12065-2919
USA

Cengage Learning products are represented in Canada by Nelson Education, Ltd.

For your course and learning solutions, visit **delmar.cengage.com**

Visit our corporate website at **cengage.com**

Printed in Canada
1 2 3 4 5 XX 10 09 08

In Loving Memory of

❧

Stan Klein

CONTENTS

PREFACE

Welcome to the second edition of *AC Theory,* which has been redesigned and updated to provide knowledge of the fundamentals to electrical technologists in apprenticeship programs, vocational-technical schools and colleges, and community colleges. The text emphasizes a solid foundation of classroom theory supported by on-the-job hands-on practice. Every project, every piece of knowledge, and every new task will be based on the experience and information acquired as each technician progresses through his or her career. This book, along with the others in this series, contains a significant portion of the material that will form the basis for success in an electrical career.

This text was developed by blending up-to-date practice with long-lived theories in an effort to help technicians learn how to better perform on the job. It is written at a level that invites further discussion beyond its pages while clearly and succinctly answering the questions of *how* and *why.* Improvements to this edition were made possible by the continued commitment by the National Joint Apprenticeship and Training Committee (NJATC) in partnership with Delmar Cengage Learning to deliver the very finest in training materials for the electrical profession.

For excellence in your electrical and telecommunications curriculum, look no further. The NJATC has been *the* source for superior electrical training for thousands of qualified men and women for over 65 years. Curriculum improvements are constant as the NJATC strives to continuously enhance the support it provides to its apprentices, journeymen, and instructors in over 285 training programs nationwide.

The efforts for continuous enhancement have produced the volume you see before you: this technically precise and academically superior edition of *AC Theory*. Using a distinctive blend of theory-based explanation partnered with hands-on accounts of what to do in the field and peppered with Technical Tips and Field Notes, this book will lead you through the study of AC Theory, from DC fundamentals, to production of AC and identification of circuit parameters, to applications of AC in generators and transformers.

K E Y   F E A T U R E S

This text has been strengthened from top to bottom with many new features and enhancements to existing content. All-new chapter features provide structure and guidance for learners. Enhanced and concrete Chapter Objectives are complemented by solid and reinforcing Chapter Summaries, Review Questions, and Practice Problems. Chapter contents are introduced at the beginning of each chapter, and then bolstered before moving on to the next chapter. Throughout each chapter, concepts are explained from their theoretical roots to their application principles, with reminders about safety, technology, professionalism, and more.

AC Theory, Second Edition, has been expanded to more fully explore a number of concepts through a major reorganization of the chapters. Combining related topics and bolstering them with additional content makes the book even more reader-friendly. As the student progresses from basic electrical fundaments learned in DC, the increased complexity of AC is added gradually to lead the student in a sequential process of adding components to create more complex circuits.

The fundamental theory of production of AC and identification of circuit parameters is used to build each additional chapter. The basic components of AC circuits are thoroughly explained; then each component is placed in a circuit to study the effects on the AC circuit. Calculations to verify circuit conditions are shown step-by-step to familiarize the students with the process and check results.

See the following pages for examples of these new features.

Running Glossaries are included in each chapter along with a comprehensive glossary at the end of the book.

High-Contrast Images give a clear and colorful view that "pops" off the page.

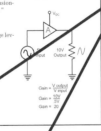

High-interest content is given in ThinkSafe!, FieldNote!, and TechTip! sidebars to make real-world connections to lessons learned in the chapters.

Sample page 115:

AMPLIFIER GAIN

An amplifier provides gain—the ratio of the output to the input. An amplifier achieves gain by converting one thing to another. For example, a lever converts force to movement or movement to force. One pound of force on the end of a beam may lift 3 pounds on the other end, but it will not lift it far. An electronic amplifier uses circuit DC power and divides that power between the output terminal and a load resistor based on the strength of the input signal. Figure 6–1 shows a block diagram of an amplifier circuit.

FIGURE 6–1 Amplifier block diagram.

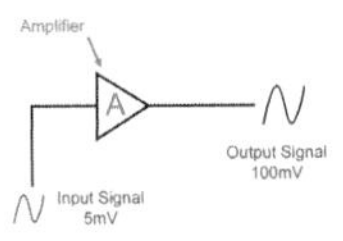

6.1 Types of Gain

If the objective for an amplifier is to produce gain, a measure of the gain of the amplifier is a measure of the amplifier's success. Gain is expressed as the ratio of output to input. Gain is shown in formulas using the symbol A with a subscript of p, i, or v to indicate power, current, or voltage, respectively.

Take care to write the result correctly; A_p, A_i, and A_v are unitless. This is because A is calculated by dividing power by power, current by current, or voltage by voltage. By not tacking units on the ends of amplification factors, statements such as "We have six times as many watts at the output than at the input" can be avoided. Normally, gain is expressed as a dimensionless number—"The amplifier has a voltage gain of 10, 20, 50, 110. . . ."

Glossary (right column):

Amplifier
A device that provides gain without much change in the original signal waveform.

Gain
The ratio of the output signal to the input signal of an **active component**.

Active component
Components of an electronic circuit that use a power source to process a signal. The processing usually involves amplification or some other change in the signal that requires additional power. BJTs, FETs, and UJTs are examples of active components.

A_p
Power gain.

A_i
Current gain.

A_v
Voltage gain.

Amplification
The process of increasing the voltage, current, or power of a signal.

FIGURE 6–2 Voltage gain.

$$Gain = \frac{V\ output}{V\ input}$$
$$Gain = \frac{10V}{5V}$$
$$Gain = 20$$

Sample page 213:

Systems and Applications   213

FIGURE 13–6 Liquid service.

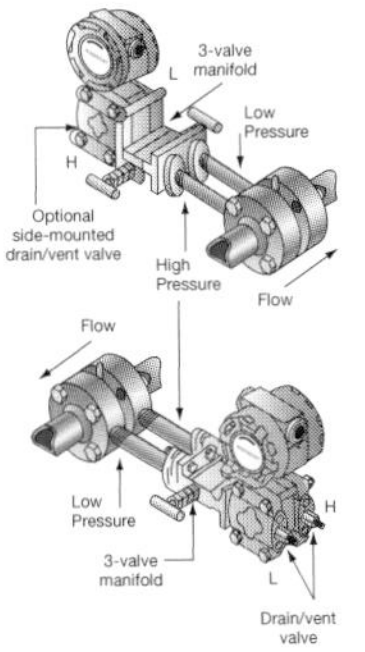

FIGURE 13–7 Steam service.

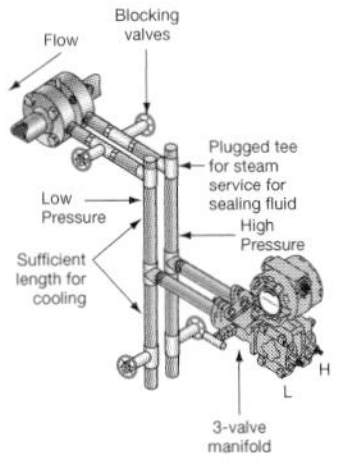

FIGURE 13–8 Gas service.

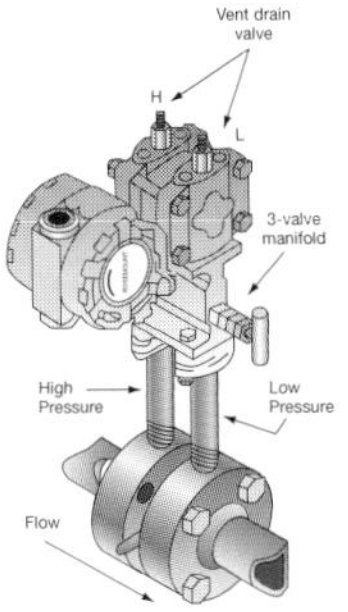

vent. If measuring gases, the valves should be mounted to allow any liquid that has collected in the process tubing to drain. If measuring steam, the valves should be mounted the same as those for liquid measurement, because the lines should be filled with water. Filling the steam service lines with water is required to prevent steam from coming into contact with the transmitter sensor (Figure 13–8). When field calibration occurs, this liquid head on the impulse lines is accounted for naturally. When bench-calibrating these devices, you must allow for the liquid head to achieve accurate calibration.

The potential zero shift due to excessive pressure is provided to the sensor of the transmitter. Elevation and suppression are two conditions that an installer must consider when measuring liquids. For special cases of gas measurement, elevation and suppression may be considerations, but we now must consider a process that is lighter than the one we are used to dealing with. Mounting details should provide all necessary requirements for mounting height with respect to tap positions. Review and follow your site-specific mounting details.

It is possible that gas-measuring devices can be mounted to show an excess or lack of process due to the relationship of the process to the mounting position. Mounting details should be provided for special service devices to ensure that they are not located improperly.

TechTip!

As you have most likely observed at a carnival or fair with large portable truck-mounted generators, as the rides start and the demand for current increases, the generators have to work harder. This is usually apparent when the diesel engines work harder, creating more noise and more exhaust. As the electrical load or power requirement diminish, the engines settle back to the idle mode or throttle back. This is a direct indication that as there is more electrical work being done, more mechanical horsepower is required. Along with the mechanical changes are the voltage regulation controls, which are not as obvious but are constantly adjusting the magnetic field to maintain voltage.

ThinkSafe!

Many pneumatic devices are spring loaded. Always use extreme caution when removing mounting bolts, because great force can be exerted as the spring tension is relieved. This force can cause serious injury and equipment damage if not planned for and alleviated.

FieldNote!

"ELI the ICE man" is a phrase used in the industry to remind practitioners of the current and voltage relationships that exist in inductive circuits represented by L and in capacitive circuits represented by C. E leads I in an inductive circuit, hence the acronym ELI. This is the same circumstance as I lagging E as we have previously learned. The acronym ICE indicates that I leads E in a capacitive circuit. Thus, "ELI the ICE man" is a mnemonic used to help memorize the relationships of E and I in various circuits. Remember that in a purely resistive circuit, the E and I are in phase and have synchronous waveforms. In AC circuits with L and C and R the current with reference to the voltage may be anywhere from 90° leading to 90° lagging, or in between.

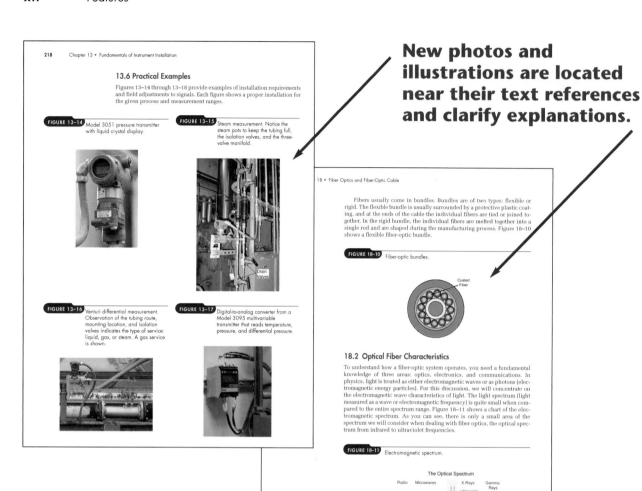

New photos and illustrations are located near their text references and clarify explanations.

End-of-chapter problems reinforce critical concepts and relate to the worked-out examples in the chapter.

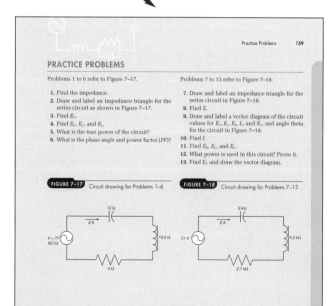

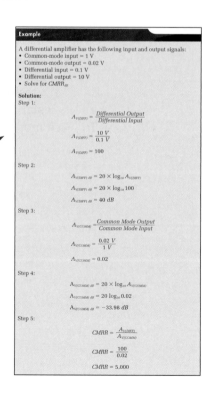

Step-by-step sample problems and solutions relate to the end of chapter exercises.

The Instructor Resource Kit is geared to provide instructors with all the tools they need in one convenient package. Instructors will find that this resource provides them with a far-reaching teaching partner that includes the following:

- PowerPoint slides (electronic and hardcopy) for each book chapter that reinforce key points and feature illustrations and photos from the book

- The Computerized Test Bank in ExamView format, which allows for test customization for evaluating student comprehension of noteworthy concepts

- An electronic version of the Instructor's Manual, with supplemental lesson plans and support

- The image library, which includes all drawings and photos from the book for the instructor's use to supplement class discussions

- A transition guide to help instructors map the changes from the previous edition to this new, stronger edition of the book

- Demonstration Exercises/Animations: a range of practical multimedia presentations and interactive simulations designed to develop and expand major concepts in electronics, including Ohm's law, amps, volts, power, series circuits, parallel circuits, series-parallel circuits, network theorems, magnetism, and electromagnetism

ABOUT THE NJATC

Should you decide on a career in the electrical industry, training provided by the International Brotherhood of Electrical Workers and the National Electrical Contractors Association (IBEW-NECA) is the most comprehensive the industry has to offer. If you are accepted into one of their local apprenticeship programs, you'll be trained for one of four career specialties: journeyman lineman, residential wireman, journeyman wireman, or telecommunications installer/technician. Most importantly, you'll be paid while you learn. To learn more, visit *http://www.njatc.org*.

A C K N O W L E D G M E N T S

NJATC ACKNOWLEDGMENTS

Principal Writer
Stan Klein, NJATC Staff

Contributing Writer
John McCord, Instructor, Vineland, NJ

Technical Editor
William R. Ball, NJATC Staff

ADDITIONAL ACKNOWLEDGMENTS

This material is continually reviewed and evaluated by Curriculum Groups who are also members of the NJATC Inside Education Committee. The invaluable input provided by these individuals allows for the development of instructional material that is of the absolute highest quality. At the time of this printing the Inside Education Committee was comprised of the following members: Chris Kelly, Chair; Richard Brooks; Robert Doustou; Carl Latona; Terry McKinch; Dan Sellers; and James Tosh.

A B O U T   T H E   A U T H O R

Jeff Keljik has been involved in electrical technical education since 1978, when he began teaching post-secondary courses for the electrical construction and maintenance program at Dunwoody College of Technology. He was head of the Electrical department at Dunwoody from 1982 to 1998. During that time, Jeff also worked with local industry and many organizations to start an Electrical Design and Management program at Dunwoody. He currently oversees the electrical installation and maintenance of the Dunwoody campus buildings.

Jeff is a member of the Education Committee and is Board of Directors-Treasurer for the North Central Electrical League. He also serves on the Education Committee for the Minnesota Electrical Association. He has co-written many courses for *NEC*® updates and exam preparation courses. Jeff has written many textbooks on power generation and delivery and on motors and controls.

Jeff has taught at many locations throughout the United States, and has helped organize programs in other countries as well as teaching classes at Dunwoody for foreign corporations.

Jeff was the recipient of the 1993 Kraemer Fellowship Award for new technology education and received awards for International Technical Teacher Education programs at Dunwoody. He was also recognized with the "good for the industry award" from industry associations.

Jeff is a graduate of Dunwoody College of Technology and has a Bachelors Degree from Metropolitan State University. He is also a member of the International Association of Electrical Inspectors. He holds Master's and Journeyman's class "A" licenses from the State of Minnesota.

Jeff lives with his wife and two children in Burnsville, Minnesota.

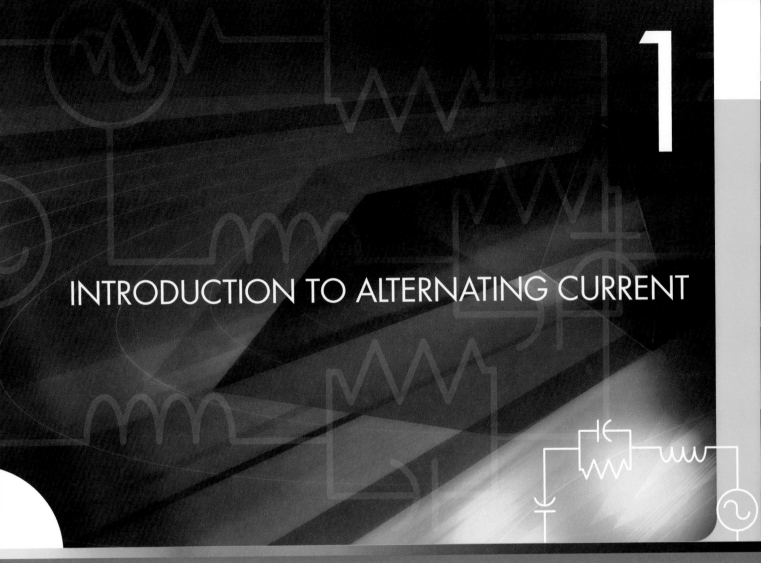

1

INTRODUCTION TO ALTERNATING CURRENT

O U T L I N E

OVERVIEW

Alternating current, usually abbreviated as AC, is used to provide over 90% of the world's electrical power requirements. In the very early years of electrical power, direct current (DC) was the most widely used form of electricity. The beginnings of the transition from a DC world to an AC world were turbulent and full of technical and political disagreements and intrigue. Thomas Edison and George Westinghouse, outspoken competitors in those early years, staked their reputations and their businesses on their particular systems.

Eventually, for a variety of technical and business reasons, Westinghouse won the dispute. His AC transformers and motors, developed by the brilliant technical genius Nikola Tesla, provided an edge over Edison's DC systems. Even so, in spite of this fairly early victory for AC, a few large American cities continued to use DC distribution in their downtown areas until the 1970s.

This brief history begins your study of AC theory. Your success in understanding the material presented in the chapters of this book is dependent on how well you understand the basic principles of DC theory. Many of the same principles that applied to DC circuitry also apply to AC circuits. However, the effects of AC create many variations on the same rules and the degree of understanding becomes more complex. The skills of a professional electrician rely on the thorough understanding of how and why AC circuits operate.

This chapter is designed to lead you from a review of DC and the theories and practices used to solve those circuit problems into the application used in AC.

OBJECTIVES

After completing this chapter you should be able to:
- Describe the difference between electron flow theory and conventional current flow theory
- Explain the difference between materials that are good conductors, good insulators, or are classified as semiconductors
- Describe various sources of electricity
- Use Ohm's law and Watt's law to explain simple circuit behavior
- Describe the effects of current flow, both DC and AC, on straight conductors
- Calculate circuit quantities on DC circuits in series, parallel, and in combinations
- Describe the electromagnetic effects that occur around straight conductors and coiled conductors

ELECTRON THEORY

Electricity is an invisible force that can produce heat, motion, light, and any number of other physical effects. This invisible "driving force" provides power for lighting, radios, motors, the heating and cooling of buildings, and many other applications. The common link among all of these applications is the electrical charge. All the materials we know—gases, liquids, and solids—contain two basic particles of electrical charge: the electron and the proton. The electron has an electrical charge with a negative polarity. The proton has an electrical charge with a positive polarity.

Atoms have three main parts: electron, proton, and neutron. The proton and neutron combine to form the atom's nucleus. Hydrogen, the simplest of all elements, has a single proton, and a single electron. Figure 1–1 is a diagram of an isotope of hydrogen called deuterium. Deuterium has one electron, one proton, and one neutron. Recall that the electron has a negative (−) charge and the proton a positive (+) charge. The neutron, as the name suggests, has no electrical charge. In other words, the neutron is electrically neutral and has no effect on the electrical characteristics of the material.

You can tell the type of an element by the number of protons in the atom's nucleus. For example, silver has 47 protons in its nucleus, iron 26, and oxygen 8. The number of protons also equals the element's atomic number in the **Periodic Table of the Elements** (see Figure 1–2). Although there are many possible ways in which electrons and protons might be grouped in atoms,

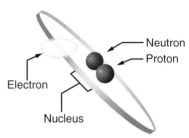

FIGURE 1-1

An isotope of hydrogen called deuterium, with one electron and one proton.

Neutron
Proton
Electron
Nucleus

Periodic Table of the Elements
A tabular arrangement of the elements according to their atomic numbers so that elements with similar properties are in the same column.

FIGURE 1-2 Periodic table of the elements.

8 8B	9 8B	10 8B	11 1B	12 2B	13 3A	14 4A	15 5A	16 6A	17 7A	18 8A
										2 **He** 4.003
					5 **B** 10.81	6 **C** 12.01	7 **N** 14.01	8 **O** 16.00	9 **F** 18.99	10 **Ne** 20.18
					13 **Al** 26.98	14 **Si** 28.09	15 **P** 30.97	16 **S** 32.07	17 **Cl** 35.45	18 **Ar** 39.95
26 **Fe** 55.84	27 **Co** 58.99	28 **Ni** 58.34	29 **Cu** 63.55	30 **Zn** 65.39	31 **Ga** 69.72	32 **Ge** 73.61	33 **As** 74.92	34 **Se** 78.96	35 **Br** 79.90	36 **Kr** 83.8
44 **Ru** 101.1	45 **Rh** 102.9	46 **Pd** 106.4	47 **Ag** 107.9	48 **Cd** 112.4	49 **In** 114.8	50 **Sn** 118.7	51 **Sb** 121.8	52 **Te** 127.6	53 **I** 126.9	54 **Xe** 131.3
76 **Os** 190.2	77 **Ir** 192.2	78 **Pt** 195.1	79 **Au** 197.0	80 **Hg** 200.6	81 **Tl** 204.4	82 **Pb** 207.2	83 **Bi** 209.0	84 **Po** 209	85 **At** 210	86 **Rn** 222

The copper atom with 29 electrons and 29 protons.

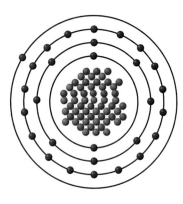

Direct current
A current that flows in one direction only without relationship to time.

Electron flow theory
The theory of current flow in which current flows from a negative charge to a positive charge; in other words, current flows from negative to positive through a circuit.

they come together in very specific combinations that produce stable arrangements (atoms). The simplest element is hydrogen and its isotopes.

Figure 1–3 shows a copper atom with a number of rings filled with electrons in orbit around the atom's nucleus. There are infinite orbital rings (shells) around a nucleus that can hold electrons, although, so far, scientists have not discovered any elements that use more than five shells. More information about the rings is described in the section "Conductors, Insulators, and Semiconductors."

1.1 Electron Flow Theory

As you studied **direct current** systems, you found that electricity is a form of energy. Movement of electrons can be generated at one location, transported to another location and then converted to another form of energy such as heat, light, sound, or mechanical motion. With DC, the electrons move through the wires in one direction. That direction is from a point of excess electrons to a point of a deficiency of electrons, in other words from a point that has too many negatively charged electrons to a point that has a deficit of electron charges or a positive charge. This book will follow the **electron flow theory,** which states that electricity (electrons) flows from negative to positive. This process is still true in AC circuitry. The major change is that the supply of electrons, or the source, continually changes electrical polarity from negative to positive and back to negative over time. This introduction of another variable (time) is a major influence on the effects of AC.

1.2 Sources of Electrical Energy

There are six methods that are known to force electrons from the valence ring of an atom to become a potential electrical current participant:

1. Friction: The shock you receive as you shuffle your feet across a wool carpet
2. Chemical: The electrical output from a battery
3. Heat: Thermocouples are used to detect flame in a furnace and control the electric gas valve
4. Pressure: Sensors are used to detect weight and produce an electrical signal
5. Light: Photoelectric cells produce electrical output based on light input
6. Magnetism: Generators use a magnetic field to create electrical power

Except for friction, electrical energy can produce the same effects as those used to create it. That is, if the proper load is connected across an electrical source, any one (or more) of the following effects will occur:

1. Chemical action: Recharging of a battery
2. Heat: Electric heat from a toaster
3. Pressure: In the ear, sound reproduction such as hearing aids
4. Light: Incandescent or fluorescent lamps
5. Magnetic fields: Motors and transformers work by electromagnetism

The chemical source of electricity is best represented in our everyday lives by the battery. The battery is a primary source of direct current (DC). Batteries are usually divided into two categories: primary cells and secondary cells. Primary cells, like the batteries used in portable radios and flashlights, cannot be recharged. Once they have depleted their chemical action, they must be thrown away. Secondary cells, like those used in automobiles, can be recharged several times.

Heat is generated whenever current flows through a wire. This happens because energy is used to move electrons. Usually, the effect of heating a wire is undesirable. For certain applications, heat can be a desirable outcome. A transfer of electrons can also take place when two dissimilar metals are joined together at a junction and then heated. This is known as the thermo-electricity process. Increases in the temperature (heat) cause a greater transfer of electrons. This type of device is called a thermocouple (Figure 1–4).

Figure 1–5 illustrates a piezoelectric device. Certain crystalline substances, when placed under pressure, generate minute electromotive forces. These forces cause the electrons to be driven out of orbit in the direction of the force. The electrons leave one side of the material and collect on the other side. Electricity derived from pressure is known as the **piezoelectric** effect. It is possible to reverse the piezoelectric process and produce pressure with electrical current. This principle is used in some applications, such as in very small earphones that use a piezoelectric crystal to produce the sound vibrations from an electrical signal.

Light energy can produce electricity, and electricity can produce light. It works both ways. Light is made up of small particles of energy called photons. When photons strike certain types of photosensitive material, they release energy into the material. There are three types of photoelectric effects

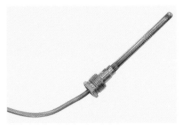

FIGURE 1–4

A thermocouple produces electricity when heated.

Piezoelectric
Generation of electricity from pressure and pressure produced by electricity.

FieldNote!

As you know, the movement of electrons from negative potential to positive potential is still called the *electron flow theory*. However, there is another way of looking at electrical flow. This original thinking, referred to as the *conventional current flow theory*, defines itself on the basis that because current flows from a positive or surplus source of current toward a point that has a negative supply of electrons, current therefore flows from positive toward negative. Many of the devices for determining current flow and magnetic field direction are exactly opposite between the two theories. When discussing circuit behavior with other people with an electrical background be sure you are using the same theories. The end results of the effects are the same, but they can be quite confusing when halfway through an analysis. Electricians tend to subscribe to the electron flow theory.

FIGURE 1–5 Piezoelectric strain gauge and microphone.

FIGURE 1–6   Photovoltaic cells produce a voltage when light strikes the device.

of interest in the study of electricity: photoemission, photovoltaic (Figure 1–6), and photoconduction.

Electrical current produces light if enough current is passed through a poor conductor. In this example, not only is heat generated, but many materials will begin to glow red or even white hot, as in an incandescent lamp. There are three other methods of producing light with electricity that do not result in as much heat loss: electroluminescence, phosphorescence, and fluorescence (Figure 1–7).

Magnetism is the primary source of electrical power. Heat and a magnetic field are generated whenever current flows through a conductor. When a conductor (a material in which the valence electrons can be easily removed from their orbit) moves through a magnetic field, electrons will move from one end of the conductor to the opposite end, creating a potential source between the two ends of the conductor. This process of producing electrical energy is known as magnetoelectricity and it makes electrical devices such as electrical magnets, motors, and transformers possible. Magnetoelectricity is the basis for commercial generators. Most electrical power is generated by this method (Figure 1–8).

FIGURE 1-7

A lamp incandesces as it is heated to produce light.

CONDUCTORS, INSULATORS, AND SEMICONDUCTORS

The transfer of electrons from atom to atom is the actual transfer of energy along a conductive path. As you studied the atomic structure of atoms you found that some elements have many electron shells and some

FIGURE 1-8   A magnetoelectric generator uses magnetism to create electricity.

have fewer. If the shells are farther from the nucleus, there is less physical attraction to the nucleus and the electron can be moved to another neighboring atom with added energy. As you saw in Figure 1–3, an atom of copper has 29 electrons. Each ring of the electron orbits has a maximum number needed to fill that ring—or shell. The innermost ring holds 2×1^2 or 2 electrons. The second ring may hold 2×2^2 or 8 electrons. The third ring may hold 2×3^2 or 18 electrons and the fourth ring could hold 2×4^2 or 32 electrons. If the rings are full of electrons the next ring begins to fill. Therefore the first three rings so far hold $2 + 8 + 18 = 28$ electrons. The 29th electron in the copper atom begins a new shell. If there is only one electron in the furthest ring, the **valence ring,** then the valence electron can move more easily as energy is added to the mix. Therefore, the best **conductors** are made of elements that have one valence electron in the valence shell. For instance, copper is a good conductor, as are silver and gold. Aluminum is also a conductor, but not as good as copper or silver, as aluminum has two valence electrons. As more electrons are in the valence shell, there is more cohesion to the nucleus and much more energy is needed to move the electrons out of their orbits. Elements with a large number of valence electrons, 8 or more, are good **insulators** of electrical flow. Glass, rubber, and most plastics are considered insulators. If enough energy is added, electrons can be made to flow through the material and the electrical insulation is said to be broken down. Materials that are in between a conductor and an insulator are known as semiconductors. **Semiconductors,** such as silicon and germanium, usually have four valence electrons and are neither good conductors nor good insulators. With the right voltage applied to the material they can conduct electricity or with absence of voltage applied they can insulate against the flow of electrons.

Valence ring
The outermost ring, shell, or orbit of electrons in an atom.

Conductor
A material whose electrons can be moved with relative ease when voltage is applied.

Insulator
A material whose electrons strongly oppose movement from one atom to the next.

Semiconductor
A material with four electrons in the valence shell. A semiconductor has more electrical resistance than a conductor but less resistance than an insulator.

AC—Alternating current
A current that varies, or "alternates," from one polarity to another. Current still flows from negative to positive but the power source changes polarity in relation to time.

Volt
The electromotive force that pushes electrons through the conductors, wires, or components of a circuit. It is similar to the pressure exerted on a system of fluid using pipes. The higher the pressure, the more flow. Specifically, the volt is the amount of work done per coulomb of charge (volts = joules per coulomb) and is represented by the letter V. In calculations, voltage is represented by the letter "E." Remember that voltage is the force required in creating flow, but volts do not flow through the circuit.

Electromotive force
The electrical pressure generated between two areas with different amounts of electrical charge. The unit of measure is the volt and represented in formulas with "E" representing electromotive force.

Ampere
The measure of a specific number of electrons is called a coulomb. That number is approximately 6.25×10^{18}, or 6.25 billion electrons. When that number of electrons passes a specific point in 1 second, or 1 coulomb per second, we say that 1 amp is flowing and is represented by the abbreviation "A". In calculations, current is represented by the letter "I" representing the intensity of flow.

OHM'S LAW

You learned Ohm's law in DC theory. The same rules apply to **AC** circuits when applying Ohm's law. There are several variations of the law that can be applied to AC circuits under various circumstances. For instance, the voltage dropped across a coil of wire with AC current flowing is only partially calculated by using a strict interpretation of Ohm's law. As we apply the rules to series and parallel circuits the same principles apply but can get a little bit more involved when we consider how frequency, or the AC waveform, plays a part in the oppositions encountered in an AC circuit.

The original Ohm's law is a law of electrical proportionality. It states that one volt will push one ampere through one ohm of resistance. Another way to look at this relationship is that the current (I) in amperes is directly proportional to the voltage (E) in volts and inversely proportional to the resistance (R) in ohms. Ohm's law can be expressed mathematically as:

$$E = I \times R \text{ or } I = \frac{E}{R} \text{ or } R = \frac{E}{I}$$

You are already familiar with these units of measure. The **volt** is the unit of electrical pressure expressed in the formula as E for **electromotive force.** The unit for current flow is expressed as **amperes** and is represented as the letter I in the formula as it was understood to represent intensity of flow. The opposition to electron flow is measured in **ohms** of resistance and is represented in the formula as R for resistance.

Example

A circuit has a voltage of 24 V and has a current of 2 A. What is the resistance of the circuit?

Solution:
Using Ohm's law, R becomes:
Given: $E = 24$ V, $I = 2$ A

$$R = \frac{E}{I} \text{ or } R = \frac{24}{2} = 12 \ \Omega$$

1.3 Cross-Sectional Area and Resistance

The cross-sectional area of a material has a great effect on the resistive characteristics of any conductor using that material. As conductors are made larger in diameter, the number of atoms contributing free electrons increases and tends to reduce the resistance for that conductor.

For example, a #6-AWG copper wire (Figure 1–9) has a diameter of 0.162 inches; therefore, the cross-sectional area is:

$$A = \pi r^2 \text{ or } \pi \left(\frac{d}{2}\right)^2 = 3.14 \times \left(\frac{0.162}{2}\right)^2 = 3.14 \times .00651 \text{ sq. in.} = .0206 \text{ sq. in}$$

However, because of the inconvenience of working with such small numbers, most calculations are done using the circular-mil area of a conductor. A mil is defined as 1/1000 of an inch, or 1000 mils equal 1 inch. To convert from wire diameters that are expressed in inches to wire diameters in mils, simply multiply the number by 1000. The #6 wire mentioned earlier would have a diameter of 0.162 × 1000, or 162 mils.

FIGURE 1-9	American wire gauge (AWG) sizes #6, #8, and #10.

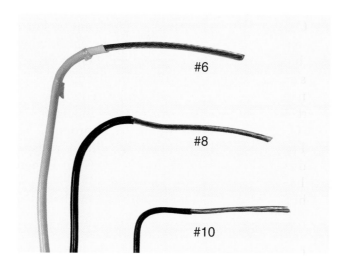

The circular-mil area of a conductor is defined as the diameter of the conductor in mils squared. The circular-mil (abbreviated as cmil) area can be found by simply taking the diameter of the wire in mils multiplied by itself. The #6 wire would have a cmil area of A = 162 × 162 = 26,244 cmils. Figure 1–10 shows two examples of cross-sectional areas and diameters for wires. Circular mils designated for various size wires can also be found in the *National Electrical Code®* book.

We will use this concept of wire resistance and the voltage drop and watt losses in a conductor later in this study of AC and AC systems.

FIGURE 1-10	The cross-sectional conducting area of round conductors is measured in circular mils.

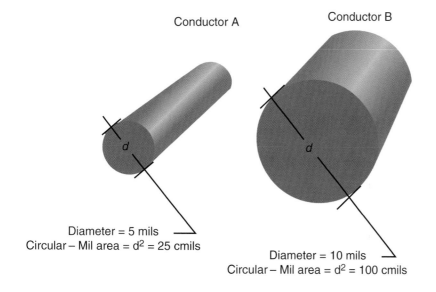

Conductor A

Conductor B

Diameter = 5 mils
Circular – Mil area = d² = 25 cmils

Diameter = 10 mils
Circular – Mil area = d² = 100 cmils

1.4 Eddy Currents

When AC flows in a conductor, the alternating current flowing through the conductor causes certain voltages to develop within the conductor. These voltages create small internal currents to flow (Figure 1–11). The name given to this current flow is eddy current as it resembles the small current that circulates in flowing water. While these internally induced currents are not flowing from one end of the conductor to the other, they are nevertheless flowing or moving internally within the conductor. Since any current flow is the result of the flow of free electrons, eddy currents are also using the free electrons to move within the conductor. Because some of the free electrons are used, fewer free electrons are available for use by the current flowing through the conductor from one end to the other. The net effect of eddy current is power loss within the conductor. This movement of electrons within the conductor uses energy and increases the overall resistance of the conductor.

FIGURE 1–11 Eddy currents circulate inside the conductors and restrict current flow.

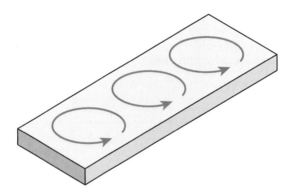

1.5 Skin Effect

In a DC circuit, the electrons travel evenly through the entire cross-section of the conductor. However, in an AC circuit conductor, besides setting up eddy currents, the voltage that creates the eddy current also causes the current flow in the conductor to be repelled away from the center of the conductor toward the outside of the conductor. The current is then forced to travel near the surface of the conductor (Figure 1–12). This effect, known as the skin effect, creates the same consequence as reducing the cross-sectional area of the conductor because the electrons are forced to flow in a smaller area concentrated near the surface of the conductor. The skin effect also causes an increase in the conductor resistance in the circuit due to power losses. Both eddy currents and the skin effect are directly related to the frequency of the circuit. Therefore, as the frequency increases, the magnitude of the eddy currents increases, causing the skin effect to also increase.

Generally, the effects of eddy currents and the skin effect do not have a critical negative impact on circuits except at higher frequencies. The use of stranded cables, which provide more surface area than a solid conductor of

TechTip!

The effects of the skin effect and eddy currents are useful to remember as they are consequences of the AC current. Because most of our calculations are for relatively short lengths of wiring at a standard AC frequency of 60 Hertz (in the U.S.), the effects are negligible and ignored for everyday wiring practices.

FIGURE 1-12 The skin effect in AC conductors causes the current to flow near the surface and not in the center of the conductor.

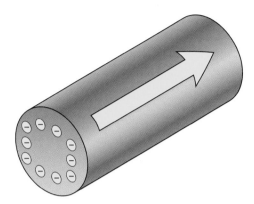

an equal size, also helps reduce the overall negative effect of these two elements within the AC circuit.

DC CIRCUIT ANALYSIS

There are three basic types of DC circuits: series, parallel, and compound (a combination of series and parallel).

1.6 Series Circuits

The total resistance in a series circuit is the sum of the individual resistors and can be calculated by the following equation:

$$R_T = R_1 + R_2 + R_3 + \ldots R_N$$

where R_T equals the total resistance and R_N equals the total number of resistors in the circuit.

Since all the series circuit current flows through every component in the circuit, each component represents a load (resistive element) in determining the total current (Figure 1–13). Ohm's law defines this relationship between voltage, resistance, and current as:

$$I_1 = I_2 = I_3 = I_T \text{ and } I_T = \frac{E_T}{R_T}$$

where I_T equals total circuit current, E_T equals total circuit voltage, and R_T equals total circuit resistance.

FIGURE 1-13 A series circuit with three resistive elements.

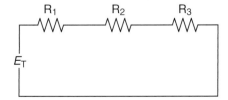

FieldNote!

If a series circuit had two 100-watt lamps connected in series, and each had the proper rated voltage applied, the total wattage of the circuit would be 200 watts. In Figure 1–14, a 60 watt rated lamp is in series with a 100 watt rated lamp, and each is designed for 120 volt operation. If this series circuit has 120 volts total applied, you will notice that the 60-watt lamp appears brighter than the 100-watt lamp. Try to explain why before reading the answer under the diagram. Section 1.7 will explain how the voltage is equal for each load in a parallel circuit (the currents are not). Therefore, if both bulbs were connected in parallel in Figure 1–14, they would both operate to their expected brightness.

The total voltage source in a series circuit is the algebraic sum of the individual voltage drops:

$$E_T = E_1 + E_2 + E_3 + \ldots E_N$$

where E_N equals the total number of voltage drops in the circuit.

Remember to keep track of the polarity of each voltage source. Voltages with opposite polarities will subtract from each other.

Like E_T and R_T in a DC circuit, the power consumed by the entire circuit is the sum of the power used by each component:

$$P_T = P_1 + P_2 + P_3 + P_4 + \ldots P_N$$

where P_T equals total power and P_N equals the total number of components in the circuit.

Example

In the circuit in Figure 1–13, the actual dissipated watts of each resistor are added together to determine total watts (or power) used in the circuit. If R_1 dissipates 50 W, R_2 dissipates 25 W, and R_3 dissipates 15 W, the total power dissipated is 90 W.

$$P_T = P_1 + P_2 + P_3$$

FIGURE 1–14   Two incandescent lamps of different rated watts connected in series.

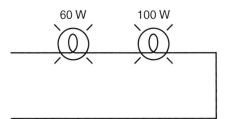

60 W   100 W

The 60 W lamp has more resistance at 120 V (240 Ω) than does the 100 W lamp (144 Ω). Therefore it takes a bigger percentage of the 120 series voltage and appears closer to its expected brightness than does the 100 W lamp.

1.7 Parallel Circuits

Parallel circuits are circuits that have more than one path for current to flow (Figure 1–15). These different paths are called parallel branches. The voltage applied to each of the parallel branches is equal to the voltage across the branch. The sum of the branch currents is equal to the total current. Mathematically,

$$E_T = E_1 = E_2 = E_3 = \ldots E_N$$

and

$$I_T + I_1 + I_2 + I_3 + \ldots I_N$$

where N is the number of branches in the parallel circuit.

FIGURE 1-15 A parallel circuit with 3 parallel branches.

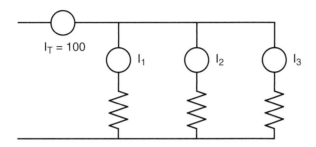

The total resistance of a parallel circuit is always less than the smallest resistance of any one branch. To calculate the total resistance in a parallel circuit, the so-called reciprocal formula is used. The reciprocal equation was developed in the DC Theory text and can take either one of two equivalent forms:

$$\frac{1}{R_T} = \frac{1}{R_1} + \frac{1}{R_2} + \frac{1}{R_3} + \frac{1}{R_N}$$

or, by rearranging the Equation algebraically,

$$R_T = \frac{1}{\left(\dfrac{1}{R_1} + \dfrac{1}{R_2} + \dfrac{1}{R_3} + \dfrac{1}{R_N}\right)}$$

where N is the branches in the parallel circuit.

Example

What is the total resistance of a parallel circuit with three branches:

$$R_1 = 50 \ \Omega, R_2 = 25 \ \Omega, \text{ and } R_3 = 15 \ \Omega?$$

Solution:
To calculate the total resistance, use the reciprocal equation:

$$R_T = \frac{1}{\left(\dfrac{1}{50} + \dfrac{1}{25} + \dfrac{1}{15}\right)} = 7.89 \ \Omega$$

Notice that the total resistance (R_T) is less than the smallest branch resistance (15 Ω).

Power calculations for parallel circuits are the same as the power calculations for series circuits. The total power in a parallel circuit is the sum of the power in the individual components of that circuit, just as the total power in a series circuit is equal to the sum of the power in the individual components of that circuit:

$$P_T = P_1 + P_2 + P_3 + \ldots P_N$$

Power calculations for these two types of circuits are also the same, in that, in both circuits, the total power is the product of the source voltage and the total circuit current:

$$P_T = E_T \times I_T$$

1.8 Combination Circuits

The first step in analyzing a combination circuit is to reduce (simplify) the circuit as much as possible. Each section to be reduced will be a group of two or more resistors with the equivalent resistance results taking the place of the group. The series resistors should be reduced first. In the example circuit shown in Figure 1–16, only one series combination exists, and it must be reduced first. Subsequently, the parallel combinations are reduced.

FIGURE 1–16   A combination circuit contains series and parallel components.

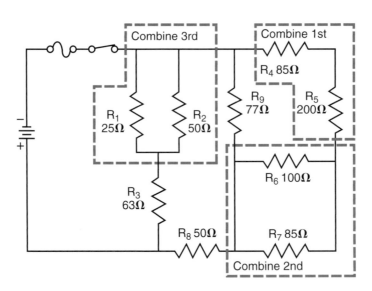

Reducing a more complex circuit to its equivalent series resistor is performed much the same as previously discussed. The general approach is as follows:

1. Identify the series and parallel current paths
2. Reduce only one part at a time
3. Ensure that all series resistors have been combined before a parallel portion is reduced
4. Combine parallel portions to a single resistor
5. Repeat combining equivalent resistors until all portions are reduced to one equivalent resistance

Once you know the total resistance in a combination circuit and the total source voltage, you can use Ohm's law to calculate the total current. However, the value of current through each respective resistor must be calculated in reverse order. Work backward from the total equivalent resistance calculated for Figure 1–16. As each equivalent branch resistance is found, apply the voltage to that branch and calculate the total branch current. Repeat the steps until all resistances, their voltage drops, and calculated currents are known.

Combine series section 1:

$$(R_4)\ 85\ \Omega + (R_5)\ 200\ \Omega = (R_{4\text{-}5})\ 285\ \Omega$$

Combine parallel section 2:

$$(R_6)\ 85\ \Omega + (R_7)\ 200\ \Omega = (R_{6\text{-}7})\ 59.65\ \Omega$$

Combine parallel section 3:

$$(R_1)\ 25\ \Omega + (R_2)\ 50\ \Omega = (R_{1\text{-}2})\ 16.67\ \Omega$$

Then combine:

$$(R_{4\text{-}5}) \text{ in series with } (R_{6\text{-}7}) = (R_{4\text{-}5\text{-}6\text{-}7})\ 344.65\ \Omega$$

Then combine $(R_{4\text{-}5\text{-}6\text{-}7})$ in parallel with (R_9)

$$(R_{4\text{-}5\text{-}6\text{-}7})\ 344.65\ \Omega \text{ in parallel with } (R_9)\ 77\ \Omega = (R_{4\text{-}5\text{-}6\text{-}7\text{-}9})\ 62.9\ \Omega$$

Combine $(R_{4\text{-}5\text{-}6\text{-}7\text{-}9})\ 62.9\ \Omega$ in series with $(R_8)\ 50\ \Omega = (R_{4\text{-}5\text{-}6\text{-}7\text{-}8\text{-}9})\ 112.9\ \Omega$

The section 3 combination of $(R_{1\text{-}2})\ 16.67\ \Omega$ in series with $(R_3)\ 63\ \Omega =$ 79.6 Ω is in parallel with $(R_{4\text{-}5\text{-}6\text{-}7\text{-}8\text{-}9})\ 112.9\ \Omega$ for a total circuit equivalent of approximately 193 Ω.

If you have an applied total voltage, you can then calculate all of the component currents, component voltages and the watts dissipated by each component.

As with other types of circuits we have studied, the total power used in a combination circuit is the sum of the power dissipated in each of the individual components in that circuit. It is also equal to the power used by the "equivalent resistance" of the circuit. If two of the three parameters of each component are known (voltage, current, and/or resistance) power can be calculated for each. Total power is then derived by adding all the individual powers:

$$P_T = P_1 + P_2 + P_3 + \ldots P_N$$

where N is the number of components in the combination circuit.

1.9 Ohm's Law and Power

The unit of electrical power is the watt. Power in a circuit is the amount of work being done per unit time. The load in a circuit uses the energy at a certain rate called power. Power, in equation form, looks like this:

$$\text{Power} = \frac{\text{Work}}{\text{Time}}$$

The work in this equation is the force applied to the circuit times the distance that the force moves something. In an electrical circuit, the power is the volts (joules/coulomb) times the amps (coulombs/second) and equals watts (joules/second). Another way to look at power is the amount of energy used per second. Mathematically, the heat or power produced can be shown as:

$$P = E \times I$$

Since $E = (I \times R)$, then substitution allows:

$$P = (I \times R) \times I \text{ or } P = I^2 \times R$$

The heat produced by the current through the resistance in the circuit is called "I^2R" losses because it is heat lost in the system. Two other formulas can be found using variations of the $P = I^2 \times R$ equation: $P = \dfrac{E^2}{R}$ or $P = E \times I$.

Example

What is the power dissipated due to heat in a motor that draws 20 A of current and has a resistance of 5 Ω?

Solution:
Since we know current and resistance, the formula to use is $P = I^2R$:

$$P = I^2R = 20^2 \times 5 = 2000 \text{ W}$$

Example

A room heater has a rating of 2000 W and uses 120 V. What is the resistance of the heater?

Solution:
Since we know the power and the voltage, the formula to use is

$$P = \frac{E^2}{R}$$

or transposing the formula $R = \dfrac{E^2}{P}$

$$R = \frac{120^2}{2000} = 7.2 \ \Omega$$

VOLTAGE DROP IN WIRE

In the DC Theory text, you learned that the voltage drop E_{VD} across a load is equal to $I \times R$ and that the resistance of wire is found by using the formula:

$$R = \frac{K \times L}{A_{cmil}}$$

where
R is the resistance of the wire in ohms
K is the resistivity of the wire in ohms per mil foot (Figure 1–17)
L is the length of wire in feet
A_{cmil} is the area of the wire in circular mils (cmil).

FIGURE 1-17 A mil-foot of wire that is 1 foot long and 1 mil in diameter will have a resistance expressed as ohms per mil-foot.

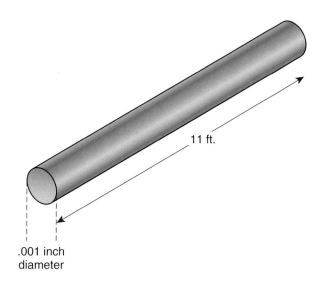

11 ft.

.001 inch
diameter

This formula shows that wire resistance is fixed and does not exist as a condition or result of voltage or current. Another variation of the formula is used to calculate the voltage drop in a single-phase circuit. This variation is:

$$E_{\text{VD}} = \frac{\text{K} \times I \times 2 \times L}{A_{\text{cmil}}}$$

Note that the multiplier of 2 takes into account that the wire makes a round-trip from the source to the load. Later you will learn to calculate the voltage drop of a three-phase circuit using a similar formula:

$$E_{\text{VD}} = \frac{(11 \times 14 \text{ A} \times 2 \times 300 \text{ ft})}{6530 \text{ cmil}} = 14.15 \text{ V dropped in the wire. (This is}$$

7.075 V to the load and 7.075 V returning from the load.)

If we started with 208 V at the breaker panel then 208 V − 14.15 V = 193.85 V at the load. This translates to $\frac{208 \text{ V}}{14.15 \text{ V}} = 14.7\%$ drop. This percent drop would be excessive and would not meet most specifications.

Example

Figure 1–18 illustrates this example. The total distance to the load is 300 feet. The wire is #12 AWG with a circular mil area of 6530 cmil and is copper with a K of 11. The load current is 14 amps. How much voltage is dropped due to wire resistance? If we started with 208 V, how much is left at the load?

Example of calculation for Figure 1–18 using aluminum conductor:

$$E_{\text{VD}} = \frac{\text{K} \times I \times 2 \times L}{A_{\text{cmil}}}$$

TechTip!

Voltage drop is nothing more than using Ohm's law and applying it to the resistance in the wire and the current carried by the wire. In other words, how much voltage does it take to get the current to the load and back to the source? Take the resistance of the wire as Resistance equals the ohms per mil-foot (the K of the particular type of wire) times the length of the wire to and from the load measured in feet, and divide by the cmils of the wire. This number is the total resistance of the circuit due to the wire. Multiply the total resistance by the total current to get the total voltage drop due to wire resistance. The formula:

$$VD = \frac{2 \times K \times I \times L}{cm}$$ takes all

these factors into account.

Reluctance

The opposition a material has to the flow of magnetic flux lines. If the material's magnetic domains are not easily aligned, the magnetic fields cannot easily pass through the material.

FIGURE 1–18 The distance to the load is 300' but the total length of conductor is 600'.

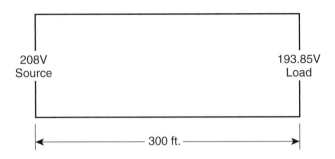

The factors stay the same except for the K of aluminum is higher than copper. Because aluminum does not conduct as well as copper, the resistivity is higher. For this example we will use a K of 18 ohms per mil-foot. The formula then looks like:

$$\frac{18 \times 14 \times 2 \times 300}{6530} = 23.15 \text{ volts dropped on the circuit of Figure 1–18}$$

PERMANENT MAGNETS

Because electricity and magnetism are so closely intertwined, a thorough understanding of magnetism is essential. You have previously studied magnetism in DC theory but a quick review can be helpful. There are three basic classifications of magnetic materials: ferromagnetic, paramagnetic, and diamagnetic. Both ferromagnetic and paramagnetic materials are metals (Figure 1–19). The magnetic properties of ferromagnetic materials allow the lines of magnetic flux to easily pass through. In ferromagnetic materials, such as iron and nickel, the lines of flux concentrate, or focus. The ability to easily concentrate these lines of flux makes the ferromagnetic materials the most easily magnetized. Paramagnetic materials, such as titanium, tend to somewhat block the passing through and concentration of the lines of flux and are therefore not magnetized as easily or as strongly. The resistance to the magnetic lines of flux is called **reluctance**. The best permanent magnets are made of steel, which is an alloy of iron and other metals. A magnet made with only soft iron does not hold its magnetism long.

The third type of material, diamagnetic material, can be made of metallic or nonmetallic materials. These materials do not allow any magnetic flux lines to pass through them but cause the magnetic lines of flux to pass around the diamagnetic material. A diamagnetic material that has a large value of reluctance is a good shield for magnetism. One example of a diamagnetic material is brass.

1.10 Electromagnetism

Whenever current flows through a material, the electron flow produces a magnetic field around the material. If the material is an electrical conductor the magnetic lines of force called flux lines emanate from the conductor

FIGURE 1–19 Ferromagnetic materials have a high magnetic ability.
Paramagnetic material has a lower magnetic ability with the
same magnetizing force.

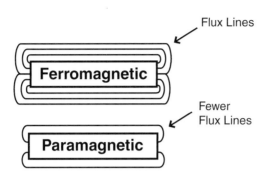

Flux Lines

Ferromagnetic

Fewer
Flux Lines

Paramagnetic

in concentric circles. The flux lines have a direction of force that is perpendicular to the conductor but do have a magnetic direction. The effect can be determined by using the left hand rule for a conductor (Figure 1–20).

Whenever current "flows" there is a magnetic field produced that is proportional to the amount of flow. The larger the current, the larger and more intense the magnetic field becomes. If the current stops, the magnetic field diminishes to zero. If the current flows in the opposite direction, then the magnetic field again builds around the conductor but with the opposite magnetic polarity (Figure 1–21).

FIGURE 1–20 The left hand rule for a conductor with the thumb in the direction of
electron flow and the fingers indicating magnetic field direction.

FIGURE 1–21 As the electrons current flow reverses, the magnetic field
changes direction.

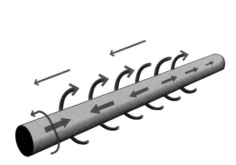

FIGURE 1–22 The X or tail feathers of the arrow indicate that current is flowing into the page in a two dimensional drawing.

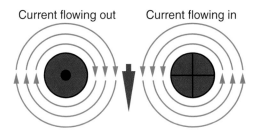

Current flowing out   Current flowing in

As you notice the cross-section of the conductors and the current direction notations in Figure 1–22, the conductor on the left has current flowing out of the page so what you see is the "head of the arrow" coming toward you. Using your left hand to grasp the conductor with the thumb in the direction of the current, the fingers indicate the direction of the magnetic field and the direction of subsequent magnetic flux is clockwise around the conductor. Conversely, the conductor on the right of Figure 1–22 has current flowing into the page so the notation is the "tail feathers of the arrow" moving away from the viewer. The left hand again indicates the concentric lines of force and the flux wraps counterclockwise around the conductor.

The left-hand rule for coils states that if you grasp a coil of wire as shown in Figure 1–23 with your fingers in the direction of the electron flow, your thumb will point to the north pole of the electromagnetic field that is created. This understanding of electromagnetic coils with magnetic fields

FIGURE 1–23 The left hand rule for a coil states that the fingers are in the direction of magnetic flux and the thumb points to the north magnetic pole.

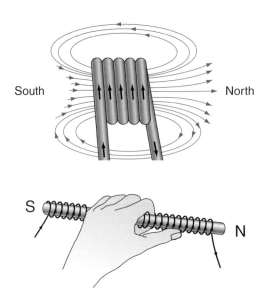

South   North

S   N

producing a known direction of magnetic flux when a known direction of current flows, will now enable you to determine the way that generators work, as explained in Chapter 2.

1.11 Magnetic Coils

If a wire is wound into a coil, as shown in Figure 1–24, and current is passed through it, the individual loops behave like small side-by-side wires with current moving in the same direction. This aiding magnetism creates one large magnetic field with a north pole and south pole like a permanent magnet. Such coils are called electromagnets.

The intensity of the field depends on two factors:

1. The number of coils—a greater number of coils will create a larger magnetic field.

2. Current magnitude—a higher current creates a larger field around the wire and thus a larger overall field.

Multiplying these two factors results in ampere-turns, which represent the total number of magnetic lines created by the electromagnet.

The effect that an electromagnet has is only partially determined by its intensity. The density of the magnetism (usually expressed as number of lines per area) is a direct measure of how "strong" the electromagnet will be. In a coil, at least two factors are important:

1. Wrapping the coils closer together or reducing the coil radius will decrease the leakage flux and increase the density of the magnetic field. Figure 1–25 shows this effect.

2. If the coil were wrapped around a core of ferromagnetic material, the number of flux lines passing through the center of the coil would increase, and so would the strength of the electromagnet. This happens because the individual molecules in the core material become polarized (aligned) in relation to the magnetic field produced by the coil.

FIGURE 1–24

The magnetic field expands around each conductor affecting the neighboring coiled wire.

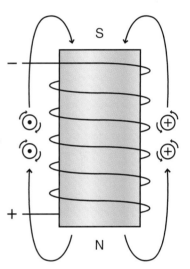

FIGURE 1–25 Changing the dimensions of the coil changes the magnetic field strength.

Original Coil with an Electromagnet Strength of 10 Ampere-Turns.

Closer (more dense) Coil (same amount of turns) with Electromagnet Strength of 30 Ampere-Turns.   Closer (more dense) Coil + More Turns with Electromagnet Strength of 300 Ampere-Turns.

TechTip!

FIGURE 1-27

Current flows when a meter is connected. Note the right end of the conductor will be the negative terminal of the generated voltage.

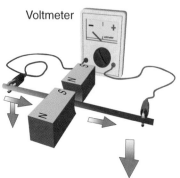

The type of core material is also a major consideration of the strength of the electromagnet. If the material has high permeability, the core allows magnetic lines of force to pass easily through the material. The area of the core material, measured as the cross sectional area of the core face at the ends, also affects the amount of magnetic flux density at the poles. As the core area increases, the magnetic flux increases. As the length of the core increase without changing the number of windings, the magnetic effect decreases. Therefore, changing the shape of the core affects the total magnetic influence of the electromagnetic core.

MAGNETISM, CONDUCTORS, AND MOVEMENT

Conductors have magnetic fields produced as current moves through them. As a conductor moves through a magnetic field, the magnetic induction process will cause a difference of electrical potential between opposite ends of the conductor and current will flow if there is a complete circuit (Figure 1–26).

FIGURE 1-26 The direction of current flow from the induced EMF as the conductor is moved through a magnetic field.

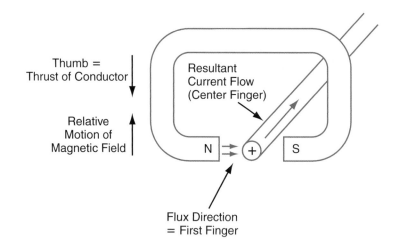

Just as more current causes a larger magnetic field in a conductor, a stronger magnetic field would cause a greater current flow in a conductor. The rate at which a conductor passes through or "cuts" a magnetic field will also determine the amount of voltage difference that is induced. Figure 1–27 is another view of the direction of current flow when a conductor cuts through a magnetic field.

The left hand rule for generators can easily predict the direction that current would flow if the direction of the magnetic field and the direction of motion are known (Figure 1–28).

Left hand rule for generators using the electron flow theory.

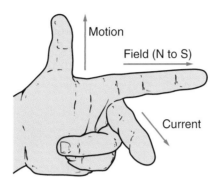

The thumb points in the direction of movement, the first finger indicates the direction of magnetic flux (from north to south), and the direction of current is shown by the center finger.

1.12 AC and DC Generators

Take a look at Figure 1–29. This simple alternating current (AC) generator is constructed of a permanent magnet, slip rings, brushes, and a single loop (armature). The armature is the rotating part of the machine in which the voltage is induced. The magnets set up the required magnetic field. As the loop rotates, each side of the loop cuts lines of flux at the same time. As the lines of flux are cut, a voltage is produced at the ends of the loop. The voltage is connected to the voltmeter by a set of sliding contacts (the slip rings). A slip ring is attached to each end of the loop.

An AC generator with armature connections through slip rings.

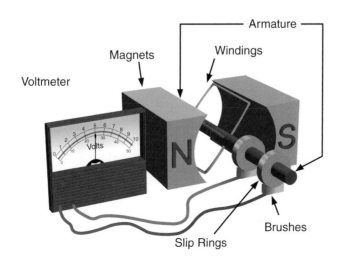

DC generators work basically the same as AC generators, except the voltage is removed from the armature by a commutator instead of slip rings (refer to Figure 1–30; see DC fundamentals for further explanation). Each commutator segment is connected to the end of one side of each loop in the generator. As the commutator segment aligns with each brush, the current flow is always in the same direction. The left brush in Figure 1–30 passes the current from the left side of the loop, and the right brush passes the current from the right side of the loop. This means that the left brush is always negative and the right brush is always positive, even though the loops change polarity as they rotate.

FIGURE 1–30 A DC generator has the same generator principles but the armature is connected to the load through a commutator.

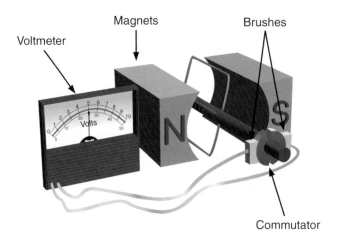

SUMMARY

The operating principles and calculations required for AC circuits require a thorough understanding of DC fundamentals. This chapter reacquainted you with the concepts we will build on in this book. Electron theory reviewed how we can get electrons to move in their orbits and the electron flow theory defined direction of movement. The energy or force needed to move the electrons was described in the "Sources of Electrical Energy" section. The differences between insulators, conductors, and semiconductors were defined to allow you to analyze the differences and help you to understand that many materials can conduct if enough electrical pressure is applied.

Conductors do have resistance to the flow of electrons and the opposition can take many forms. Wire resistance was discussed as well as skin effect and eddy current losses. A review of DC circuit analysis was designed to remind you of all the rules for DC as most will also apply to AC and some need to be modified. Permanent magnets and electromagnets were also reviewed as a critical part of the AC circuit theory and operations. The basic theory of generation was presented to link the DC generator to the AC generator and as an introduction to subsequent chapters.

REVIEW QUESTIONS

1. Describe how good electrical conductors are different from insulators and semiconductors.
2. How is the resistance of a conductor related to its diameter, its cross-sectional area, and its resistivity?
3. List four forms of electricity production.
4. Describe Ohm's law.
5. What is the effect of "skin effect" in electrical circuits?
6. What is the left hand rule for generators? How can it be used to determine the polarity of a generator voltage output?

PRACTICE PROBLEMS

1. Look at Figure 1–31 and assume that the power supply is a 120 V battery. Calculate the following:
   a. Total circuit current
   b. Total circuit power
   c. Voltage drop, current flow, and power dissipation for each of the five resistors

FIGURE 1–31   A complex circuit with five resistors.

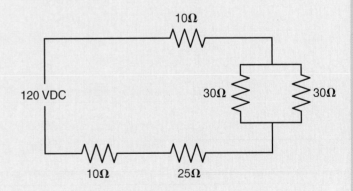

2. Refer to Figure 1–32. What is the magnetic polarity of the right end of the magnetic coil with the circuit polarity as shown?

3. Using the information in Figure 1–33, calculate the voltage drop between the source and the load.

4. Explain what is meant by circular mils.

5. What is the effect of eddy currents on the current flow in a conductor?

FIGURE 1–32 An electromagnetic coil with polarity shown.

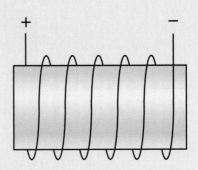

FIGURE 1–33 A typical wire resistance circuit.

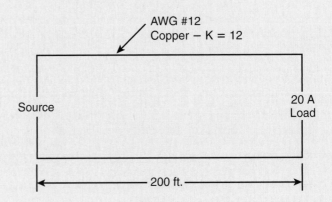

2

PRODUCTION OF THE AC SINE WAVE

OVERVIEW

The production of AC is a relatively simple concept, but to understand the process and the results of the AC generation is a more complex study. In the production of electricity the need for motion is usually accomplished by a mechanically driven system (Prime mover). Some sort of energy is needed to create the voltage. This is accomplished in large systems by steam turbines. The steam is developed by burning coal, or other fossil fuels, or by creating steam by nuclear fission. On smaller scale generators the mechanical motion (or Prime mover) is provided by gasoline or diesel engines. There are other methods too, but these are the most common. Electricians are usually not concerned with the mechanical input power, but with the electricity that is generated. AC has many new terms and conditions that are not studied in DC theory.

OBJECTIVES

After completing this chapter you should be able to:
- Describe the factors needed to generate a voltage
- Name the components of the AC sine wave
- Calculate the instantaneous values, RMS, peak amplitude, peak-to-peak, the period and frequency, and average values of AC
- Determine the synchronous speed of a generator to produce a desired frequency
- Calculate the number of poles of a generator given the speed and the frequency
- Explain the relationship between mechanical degrees, electrical degrees, and number of poles

AC GENERATORS

Generation of AC is the most widely used form of production of electricity. The concept of electromagnetic mechanical generation of power is a simple one. The items needed to generate voltage are a conductor, a magnetic field, and relative motion between the conductor and the magnetic field. AC generators, or alternators, are energy converters. They change mechanical energy from the prime mover to electrical energy. The opposite conversion may take place at the other end of the circuit where a motor converts electrical energy back to mechanical energy. A simple AC generator is shown in Figure 2–1. This example illustrates the three components of generation. There is a magnetic field, which has a magnetic flux traveling from north to south magnetic poles.

FIGURE 2–1 Simple rotating armature AC generator with slip rings.

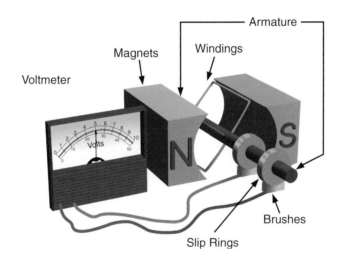

Figure 2–2 demonstrates the left hand rule for generation in electron flow theory analysis. The <u>F</u>irst finger of the left hand represents the direction of the magnetic <u>F</u>lux, again pointing toward the south magnetic pole. As the conductor moves across the face of the south pole, the conductor <u>T</u>hrust upward is represented by the <u>T</u>humb of the left hand. The result of the movement (cutting magnetic flux) causes a displacement of the electrons in the conductor, creating a surplus at one end and a deficiency at the other end.

In other words, a difference in potential or a voltage difference is created. If there is a circuit for current to flow such as shown by the meter, electron current will flow from the negative to the positive terminal in the connected circuit as would be indicated by the <u>C</u>enter finger. Be aware that this is the external circuit and that internally to the generator it looks like the current is flowing to the negative terminal. We will need this information later as we analyze other circuits. By following the same conductor to the opposite side of the generator you will notice that the direction for the thrust is now

FIGURE 2–2

Left hand rule for generators is used with electron flow theory.

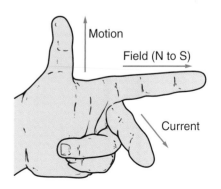

downward. By positioning your left hand correctly, you will notice the direction of the current flow is reversed. Sometimes this is easier to see in a two-dimensional drawing as in Figure 2–3. Because the external circuit is still connected to the same end of the same conductor through the slip ring, the current flow in the external circuit also reverses.

Figure 2–3 uses the notation of ✘ and • to indicate current flowing into the coil conductor and out of the coil conductor, respectively. The ✘ symbol represents current going into the coil conductor and the • symbol indicates current is coming out of the coil conductor. You can see that every time one side of the coil passes under one pole it has one electrical polarity, and creates an opposite polarity when it crosses the opposite magnetic pole. This two-pole generator has voltage induced into the spinning conductor (known as the **armature**) whenever it cuts a magnetic field. When the conductor cuts through 10 million lines of magnetic flux in 1 second it will produce 1 volt as shown in the formula: 10^6 magnetic lines per second = 1 volt.

Armature

The component in the process of generation that has voltage induced into it. It is not always the moving component but is the place where generated voltage is collected.

FIGURE 2–3   Direction of electron flow expressed in two-dimensional representation.

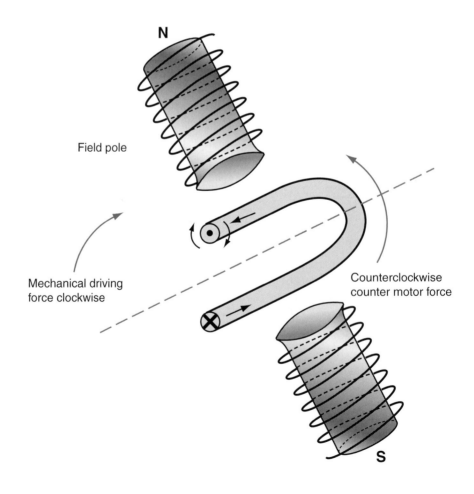

The number of lines cut depends on the strength of the electromagnetic flux. As you notice, the speed of the cutting is also a factor in the amount of voltage induced in one conductor.

Figure 2–4 shows a conductor spinning in a circular motion. If we just refer to one end of the spinning conductor at 0° of rotation there is no cutting action and no voltage induced. The graph of this is shown at the right. As the conductor spins clockwise we begin to cut flux lines and the voltage graph increases with time. As the conductor is directly under the north pole it is cutting the maximum lines of flux and it is 90° through its rotation. Continuing through the rotation to 180° of rotation we again are traveling parallel to the lines of flux and no cutting action takes place, so again, we have no voltage as indicated at 180° on the corresponding graph. At 270° of rotation we are cutting the maximum amount of flux but the current in the conductor is reversed as indicated by the graphical representation. As the conductor travels parallel to the magnetic field, as it does at the left and right of the arc, it induces no voltage, and the output voltage goes to zero. Thus, a waveform with a maximum and a zero point is achieved as the conductor spins in a circle.

FieldNote!

The oscilloscope is a tool that provides visual representation of a waveform. Through the amplitude controls, we can measure the various components of the waveform such as peak or peak-to-peak. The scope allows us to adjust the time in order to see small factions of a second so that we can actually view 60 cycles per second or more with relative ease. The oscilloscope can be thought of as a voltmeter but we can actually see the waveform displayed on the screen.

FIGURE 2–4   Mechanical generation of sine wave and resultant waveform.

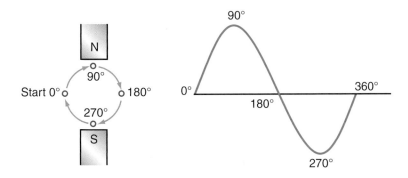

FIGURE 2–5

Photo of oscilloscope view of 60 Hz sine wave.

A sine wave is the result of a spinning conductor within a magnetic field. If you use the sine of the angle of movement and multiply it by the expected maximum voltage then the resultant is plotted on a graph, and for each of the 360° of rotation you will see a figure known as an AC sine wave. As indicated by Figure 2–5 the sine wave for the generated voltage is constantly changing from zero to maximum and back to zero. It then reverses direction of voltage and increases to maximum negative peak before again returning back to zero.

This full rotation took 360° of mechanical rotation and because it is the sine of those mechanical degrees, the electrical wave follows the same degrees but is expressed as electrical degrees. One cycle of voltage from zero to positive, zero to negative, back to zero voltage is one cycle of AC voltage. If we can spin the conductor one revolution

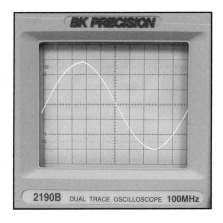

Frequency
The number of complete wave-forms completed in 1 second. Frequency is a measure of how often the waveform is completed in relation to time, expressed in cycles per second. The unit of measure is the hertz (Hz) named after Heinrich Hertz.

in one second, we record one cycle per second (Figure 2–6). The measure of how frequently we can produce a sine wave is called the **frequency** of the waveform. If we can get 60 cycles in a second, then we record 60 cycles per second (CPS). The *CPS* is an older term and you may see it on old equipment. *Hertz* is the new term for the cycles per second unit of measure (60 CPS = 60 hertz, or Hz). In some instances we refer to the real time it takes for a waveform to complete a full cycle—360° as a period. The period is calculated by taking the inverse of the frequency in Hz. The period of a 60 Hz waveform is $\frac{1}{60}$, or .0166 seconds. This tells us exactly how much time elapses as the AC oscillates completely through its two half cycles.

FIGURE 2–6 A plot of all instantaneous values creates a sine wave.

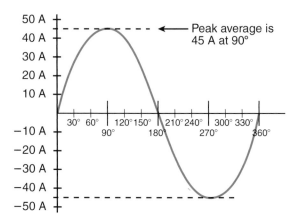

There are different ways to measure AC and designations given to various parts of the waveform. Most times we will not be concerned with the actual waveform but rather on its effects in the electrical circuits. As you study motor controls and electronic systems you will need to be aware of the specific waveforms and how we use them to provide different functions.

INSTANTANEOUS VALUE OF A SINE WAVE

As you learned earlier, the AC waveforms used in electrical power applications are in the form of sine waves. This means that any instantaneous value on the wave can be determined by using the formula:

$$I_{inst} = I_{peak} \times \sin(\theta)$$

$$E_{inst} = E_{peak} \times \sin(\theta)$$

where I_{inst} is the instantaneous value of the current at any point during the waveform

I_{peak} is the maximum value of current obtained during any point of the waveform

$\sin(\theta)$ is the sine function of any angle "theta" represented by the degrees of waveform

Refer to Figure 2–6. You can find the instantaneous value by multiplying the peak voltage, or current, times the sine of the angle at this specific point (a specific point on the sine wave at a specific point in time) at which you want the measurement. The formula for this calculation is given next.

Example

What is the instantaneous current value of a 45 ampere (peak) AC circuit at 30°, 45°, and 60°?

Solution:

$$I_{\text{inst}} = 45 \times \sin (30°) = 45 \times 0.5 = 22.5 \text{ V}$$

$$I_{\text{inst}} = 45 \times \sin (45°) = 45 \times 0.707 = 31.8 \text{ V}$$

$$I_{\text{inst}} = 45 \times \sin (60°) = 45 \times 0.866 = 38.97 \text{ V}$$

AC SINE WAVE MEASUREMENTS

There are four designations used to measure current and voltage associated with AC sine waves: peak, peak-to-peak, RMS, and average. These values are used at various points in your study of AC.

2.1 Peak and Peak-to-Peak

The peak value is measured from zero to the highest value obtained in either the positive or the negative direction. Therefore, the peak-to-peak value is twice that of the peak because it is measured from one peak value to the other. Particularly in the study of electronic control of AC power, the **peak-to-peak** or (P-to-P) values become important. The electronics actually have to be sized to handle the full peak value in one direction and have to be able to block the full peak value in the opposite direction. The peak-to-peak value is the real waveform available at the typical receptacle outlet.

The sine wave creates a shape that is not a semicircle so the typical circle or half circle calculations do not provide a true equation of the affect of the sine wave. The shape of the wave creates a condition where the peak value of the voltage waveform can be calculated by using the effective value of the voltage times the square root of two ($\sqrt{2} = 1.414$). If we measure the RMS value of 120VAC at the receptacle, we know the actual sine wave peak is 120V $\times$ 1.414 or 169V. This takes into account the sine wave shape and allows us to calculate the actual values if we were to plot all the individual voltage points along the waveform. The reciprocal of 1.414 is .707. This allows us to multiply the peak value by .707 to determine the RMS value.

To determine the full sine wave value from peak to peak, multiply the peak value by 2.

Peak-to-Peak
The full value of the AC sine wave measured from the positive peak to the negative peak is referred to as the P-to-P value.

FieldNote!

Measuring 120 V with the standard meter really means the peak is nearly 170 V and the P-to-P is nearly 340 V (Figure 2–7).

FIGURE 2–7   120 V RMS on the meter is actually approximately 340 V peak to peak.

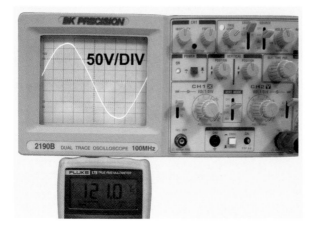

2.2 RMS

The RMS value is also called the effective value, or E_{eff}. It is so called because the AC waveform has peaks and valleys but the effective value of the waveform will have the same heating effect as a DC voltage of the same value. Consider Figure 2–8. The blue trace (trace 1) is a DC voltage

FIGURE 2–8   The RMS value or effective value is .707 times the peak.

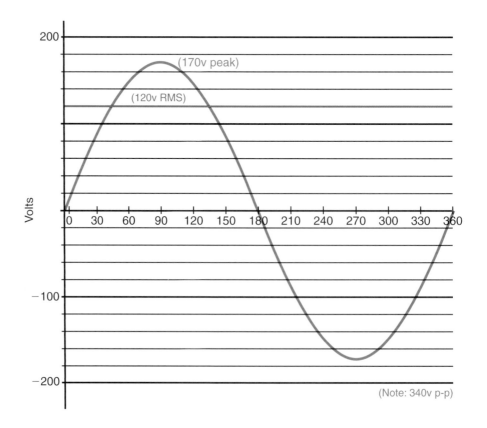

equal to 120 V. If the effective voltage of 120 V was applied to a 100-Ω resistor, the power dissipated could be calculated using the familiar power formula:

$$P = \frac{E^2}{R} = \frac{120^2}{100} = \frac{14,400}{100} = 144 \text{ W}$$

What is the peak value of an AC sine wave that will dissipate the same amount of power 120 VDC? Imagine connecting an AC voltage source and a wattmeter to the 100-Ω resistor. Now, gradually increase the AC voltage until the same 144 W are being dissipated. For a sine wave, this will occur when the peak voltage is at $120 \times \sqrt{2} = 169.71$ V. Put another way, 120 V is the effective value of an AC voltage with a peak of 169.71 V.

This value can be calculated mathematically by squaring each instantaneous point on the AC waveform, then taking the mean (average) of all the squared values and finding the square root of the average, which is the root of the mean of the squares or Root Mean Square **(RMS)** of the AC waveform. This is where the term *RMS* originates. For a sine wave, the RMS value is equal to the peak value multiplied then divided by the square root of 2 or 1.414.

Example

A circuit has a peak value of 423 amps. Find the RMS value.

Solution:

$$I_{RMS} = \frac{I_{PEAK}}{\sqrt{2}}$$

$$I_{RMS} = \frac{423 \text{ A}}{1.414} = 299 \text{ A}$$

or

$$I_{RMS} = 423 \text{ A} \times .707 = 299 \text{ A}$$

A circuit has an RMS value of 115 amperes. Find the peak value.

Solution:

$$I_{PEAK} = I_{RMS} \times \sqrt{2} = 115 \text{ A} \times 1.414 = 162.6 \text{ A}$$

2.3 Average

The average value does not apply to a pure AC sine wave. It is used primarily when an AC sine wave is changed to a rippling DC waveform by a full-wave rectifier. Recall that the effective value causes the same amount of heat throughout both the positive and the negative half cycles as the same value of DC. The average value is the actual average of all the instantaneous voltages or currents values across a full cycle. This is the reason that a pure sine wave has no DC value; it is above the zero exactly as much as it is below. Therefore, a pure AC sine wave's average is zero.

A rectified sine wave, on the other hand, has most or all of its waveform above zero. The actual DC value depends on whether it is a full-wave rectifier (Figure 2–9) or a half-wave rectifier (Figure 2–10). As you note in the figures the electron flow flows against the arrow symbols of the diode rectifiers. This is a result of the differences between electron flow theory and the conventional flow theory. As you look at the two figures of Figure 2–9, the arrows indicate the two directions of the AC applied voltage, looking at them as one voltage polarity at a time. The input to the full wave rectifier is an AC waveform, but the output to the load is a pulsating DC waveform.

FIGURE 2-9   The average value is the voltage from a full wave rectified DC.

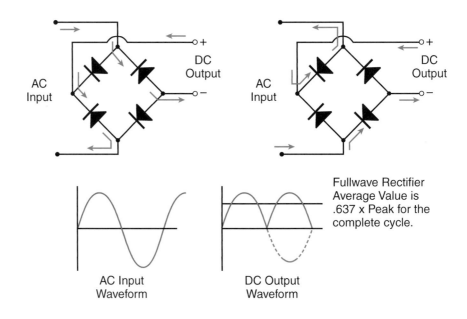

FIGURE 2-10   The average value used with a half wave rectified DC.

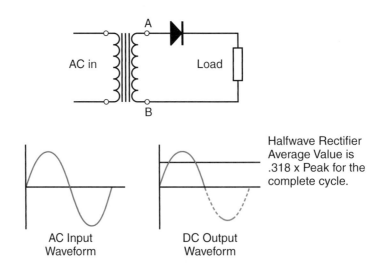

FIGURE 2–11   An assortment of photos of variable frequency drives (VFD).

We have covered formulas relating to the sine wave here. The calculations for half-wave and full-wave rectifiers will be covered in more detail in the Electronics text.

The formula for calculating the average value of a full-wave rectified wave is calculated mathematically as:

$$V_{AVE} = \frac{2 \times V_{peak}}{\pi} = \frac{2 \times \sqrt{2} \times V_{RMS}}{\pi}$$

Example

The effective value of a full-wave rectified sine wave is 30 A. Find the average value.

Solution:

$$\frac{2 \times V_{RMS} \times \sqrt{2}}{\pi} = \frac{\sqrt{2} \times 60}{\pi} = 27 A$$

If you were to surmise that the average value of a half-wave rectified sine wave would be equal to half the value of a full-wave rectified sine wave you would be right. The formula for calculating the average value of a half-wave rectified sine wave is:

$$V_{AVE} = \frac{V_{peak}}{\pi} \text{ or } \frac{V_{RMS} \times \sqrt{2}}{\pi} \text{ or } V_{RMS} \times .450$$

FieldNote!

Note that a shortcut can be used for this formula based on the fact that $\frac{(2 \times \sqrt{2})}{\pi} = 0.9$. Thus, multiplying the RMS value of the full-wave rectified sine wave by 0.9 gives the average value.

CONSTRUCTION OF AC GENERATORS

AC generators are designed in two styles. One style, the revolving arma-ture, is the easiest to visualize and the easiest to explain. As was seen in the previous section, the loop spinning in the magnetic field has a voltage induced into it. The loop that has voltage induced into it is referred to as the armature. The amount of voltage induced in one wire is small, cutting 10 million lines of flux per second, the amount of voltage generated can be increased if we add the effects of many conductors together. In effect we place them all in series and voltage induced in each conductor is added to the circuit. To do this the conductor is wrapped in a coil so that many con-ductors are traveling through the magnetic field together. The more "turns" of the coil the larger the voltage. This is a design feature of the generator and is not easily altered as the generator is in use. As was previously ex-plained, if the magnetic field is more dense, the more lines of flux are cut and the voltage induced is increased. There is a limit to how strong the magnetic field can be.

Another factor previously discussed is the speed of the cutting action. The more flux lines cut per second, the higher the voltage. Assuming the flux density stays the same, speed can affect the voltage. This factor of speed creates a problem for AC generation. The speed of the rotating arma-ture is also directly related to how many complete sine waves are produced in one second. The hertz of the output voltage is affected. Remember one revolution per second for a two-pole generator creates one Hz. A speed of 3600 revolutions per minute (rpm) for a two-pole generator is the same as 60 revolutions per second; 60 revolutions per second for a two-pole gener-ator produces the standard US frequency of 60 Hz.

The formula for determining the exact speed that a generator must spin to create a desired frequency is the **synchronous speed** formula:

Synchronous speed
The exact speed of a rotating member to produce a desired frequency for a generator. This same term is used to describe the speed that is produced by a specific frequency to make an induction motor run.

$$\text{Synchronous speed} = \frac{(120 \times \text{Frequency})}{\text{Number of poles}}$$

- Where synchronous speed is the exact speed a generator must spin to develop the desired frequency output
- 120 is a constant that results from converting cycles per second to rpm—a factor of 60, and converting number of poles from actual pole faces to pairs of poles—a factor of 2—yielding a constant conversion of 120
- Frequency is measured in Hz
- Number of poles is always an even number

REVOLVING ARMATURE AC GENERATORS

Revolving armature AC generator
A generator that has the field windings on the stator and the armature windings on the rotor.

The **revolving armature AC generator** is not as common as the revolving field generator. It is similar in construction to the DC generator in that both types use some form of mechanical device to connect the external circuit to the armature circuit. The AC generator uses slip rings instead of a commu-tator. The purpose of the slip rings is to directly connect the armature wind-ings to the load. Since no commutator is used, the output is an alternating voltage instead of a direct voltage.

FIGURE 2-12 A revolving armature generator with brushes and slip rings.

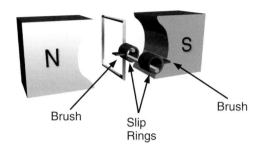

Brush   Slip
Rings   Brush

The slip ring, unlike the segmented DC generator commutator, is constructed as a continuous ring and is not broken up into isolated segments. Each individual end of the rotating loop of the armature is connected to a separate **slip ring** (Figure 2–12).

Although the revolving armature AC generator can produce an alternating output voltage, the slip rings and brushes limit the size of load it can carry. The higher currents require large slip rings and large brushes and highly increased maintenance. The high voltages required by some loads mean that much heavier insulation must be used. Taken together, these features limit the kilovolt-ampere (kVA) capacity of the revolving armature generator.

Slip rings
Continuous bands of metal installed around a shaft. Slip rings are connected to the rotating windings and provide a path for the current to reach the brushes.

ALTERNATOR CONSTRUCTION

Another more practical alternator is the **revolving field generator**. This approach allows the magnetic field to spin and the coils of wire where voltage is induced—still in the armature—to remain firmly attached to the stationary part of the generator known as the *stator*. This means that the slip rings we used to connect to the circuit in the revolving armature-type generator are no longer needed to collect the induced voltage. These sliding contacts on the slip rings are a point of concern. If we can avoid the use of a slip ring to connect to the load circuit we can eliminate many problems.

The same three factors are needed for the generator. The conductor is coiled and is now placed in the stator. The magnetic field is placed on the rotating component (the *rotor*) and there is still relative motion between the two.

Revolving field generator
A generator that has the field windings on the rotor and the armature windings on the stator.

REVOLVING FIELD AC GENERATOR

The same synchronous speed formula holds true whether the magnetic field rotates or the armature rotates. The same sine wave is produced as the sine of the angle of mechanical rotor movement is used to plot the AC waveform from zero, to peak, to zero, to negative peak, and back to zero. Figure 2–13 shows a two-pole rotating field alternator.

In small-scale generators the magnetic fields we have discussed that may be mounted on the stator or the rotor could be a

FIGURE 2-13

A revolving field AC generator with permanent armature connections.

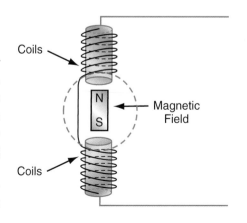

Coils

Magnetic
Field

Coils

PMG
Abbreviation for permanent magnet generator. In this type of generator, the field is provided by a very strong permanent magnet.

permanent magnet. These generators are referred to as a **permanent magnet generator (PMG).** For most large commercial and electric utility generators the magnetic field will be an electromagnet. As you know, the strength of the electromagnet can be changed by adjusting the amount of amperage that flows through the coil. The electromagnet is powered by DC so the magnetic polarity of the magnetic field does not change. If the generator is a revolving armature, then the DC is directly connected to the stator windings to form a stationary magnetic field as we previously saw. If the magnetic field is on the rotor, as in the revolving field generator, then DC must be supplied to the rotating member. This may be done with slip rings or other means that will be discussed later. It may be difficult at first glance to determine if an alternator is a revolving field or a revolving armature.

GENERATOR EXCITATION

Excitation current
The current supplied to the field of an AC generator. The excitation current creates the magnetic field that the armature cuts through.

We have now generated a sine wave, with a specific frequency based on the synchronous speed and the number of poles. We can change the voltage by changing the quantity of flux lines cut by the armature. To do this we need to control the DC current to the DC electromagnet. This DC electromagnet, whether spinning or stationary, is referred to as the *exciter*. The amount of current, called the **excitation current,** and voltage supplied to the exciter is the excitation power. The excitation power is the DC input to a generator that causes the AC RMS voltage generated to increase or decrease because the number of conductors cannot change and the speed cannot change because of the need to maintain a desired frequency. The excitation power input of the electromagnet exciter will affect the peak-to-peak voltage, but not the frequency. Figure 2–14 shows a two-pole rotating field alternator with slip ring input.

FIGURE 2–14   DC power is supplied to the rotating field by slip rings.

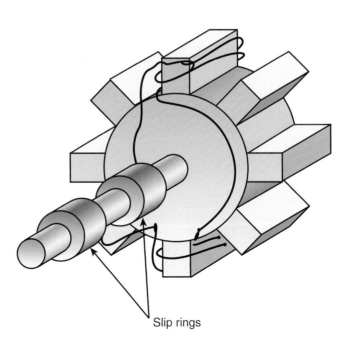

Slip rings

MULTIPLE-POLE GENERATORS

The synchronous speed of a two-pole generator is 3600 rpm to maintain 60 Hz. If we add another set of magnetic poles—north and south—and space them around the stator at the one-quarter rotation point, then spin a conductor past them, we can get a conductor to pass north and south poles with only 180° of mechanical rotation (Figure 2–15). This means the voltage will create a complete sine wave in 180° or two sine waves in 360° of mechanical rotation. The formula for electrical degrees relating to mechanical degrees is:

Electrical degrees = mechanical degree × number of pairs of poles

In this case with four magnetic poles we generate 360 electrical degrees = 180 mechanical degrees × 2 pairs of poles.

In other words we only need to spin the armature half as fast to get the same number of cycles per second. This is confirmed in the synchronous speed formula:

$$\text{Synchronous speed} = \frac{120 \times 60 \text{ Hz}}{4 \text{ poles}} : \frac{7200}{4} = 1800 \text{ rpm}$$

FIGURE 2–15 Two-pole AC generator produces 2 full sine waves for 360° of rotation.

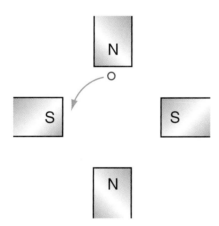

We must always have an even number of poles—a matched set of north and south. If we add two more poles for a total of six, we see that we need to move even fewer mechanical degrees to get a 360° sine wave. Or 360 electrical degrees = X mechanical degrees × 3 pairs of poles

Transposing the formula: $\dfrac{360 \text{ electrical degrees}}{3 \text{ pairs of poles}} = 120$ mechanical degrees

$\dfrac{120 \times 60 \text{ Hz}}{6 \text{ poles}} = 1200$ rpm for synchronous speed of a 6-pole generator producing 60 Hz. 1200 rpm is one-third as fast as the two-pole generator to get the same frequency.

We have now discussed the generation of single-phase AC and the resulting AC waveform. We will discuss other forms of generation and the methods to control the generator output other than frequency in later chapters.

FieldNote!

An example of a multiple-pole generator is pictured in Figure 2–16. This generator is operating in a hydroelectric generating station and is producing 60 Hz. It has 72 poles. What is the synchronous speed? 100 rpm.

FIGURE 2–16   Photo of slow speed hydroelectric generator.

SUMMARY

For the production of electricity or the force to move electrons we need three factors. We need a conductor with the electrons able to move, a magnetic field, and motion between the conductor and the magnetic field. The left hand rule for generation allows us to predict which way the electrons will move as the conductor cuts through magnetic flux lines. As the conductor is coiled and placed on a rotor, the output of the generator can be taken directly off the armature with a slip ring and the resultant voltage will oscillate between two electrical polarities. Not only does the polarity change, but the amount, or amplitude, of the voltage changes, producing a sine wave. By adjusting the speed of the rotor we can produce more or less sine waves per second, thereby changing the frequency of the waveform.

Each feature of the sine wave can be identified by the instantaneous value at any point along the waveform by using the sine of the angle. Likewise we can identify the peaks, the peak-to-peak amplitude, and the frequency and the period of the waveform. The measurement we use daily is the RMS value of the AC sine wave. The average value of the AC was introduced in this chapter as a feature used with electronic conversion of the sine wave.

Finally, in this chapter we analyzed how to produce voltage by several different styles of generators. The basic concept using a rotating armature was expanded to a stationary armature–rotating field generator where the magnetic field spins and the electricity comes off solid connections to the armature on the stator. The concept of multi-poled generators was explained as a way to get the same frequency of the waveform, although it needs a slower synchronous speed. This changes the relationship between mechanical degrees of rotation and the number of electrical degrees on the waveform. The waveform stays at 360° but the mechanical generator no longer needs to spin 360 mechanical degrees.

REVIEW QUESTIONS

1. Describe how to determine direction of electron flow when a conductor is moved through a magnetic field.

2. In order to produce 60 Hz from a four-pole generator, what is the synchronous speed?

3. What peak-to-peak voltage is required to yield a 100 V RMS with an AC generator?

4. What is an advantage of a revolving field generator?

5. What is the actual time that elapses for one cycle of a 50 Hz waveform?

6. Calculate the instantaneous value of current at 30° if the peak current is 100 A on an AC sine wave.

7. What factor do you use to multiply the peak voltage to get RMS voltage?

8. Describe RMS value of voltage.

9. Describe how to determine the number of electrical degrees of a waveform by determining how many poles a generator has.

10. What is the peak value of a sine wave that measures 120 V AC RMS?

PRACTICE PROBLEMS

Use Figure 2–17 to answer the following questions.

1. If the peak of the sine wave is 141.4 volts, find the RMS voltage value.

2. If the effective value of the AC sine wave is 50 volts, find the peak value.

3. Find the peak to peak voltage of the AC sinewave if the RMS value is 230 V.

4. If the second half of the waveform is inverted and appears above the zero reference line, the waveform is called _____.

5. If the peak-to-peak value of the AC waveform is approximately 340 V, find the AC RMS voltage.

6. If the waveform takes $\frac{1}{60}$ of a second to complete a full sine wave, what is the frequency?

7. What is the relationship between the numbers 1.414 and .707?

8. Explain why the DC average is lower for a half wave rectifier circuit, when compared to a full wave rectifier circuit.

9. Assume the waveform is a current waveform. IF the effective AC current is 15 A, what is the P-to-P current?

10. At what degree points on the waveform do peak values occur?

FIGURE 2–17 The RMS value or effective value is .707 times the peak.

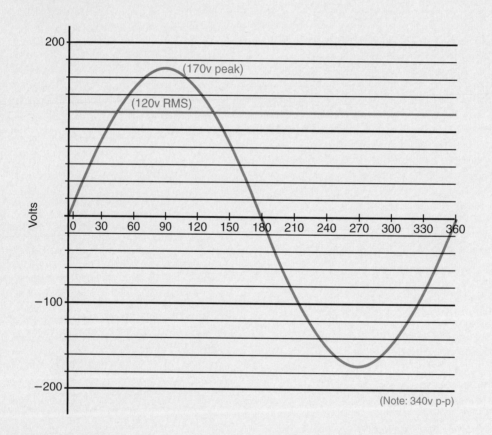

(170v peak)

(120v RMS)

(Note: 340v p-p)

3

INDUCTORS, SELF-INDUCTION, INDUCTIVE REACTANCE, AND IMPEDANCE

OVERVIEW

The previous chapter introduced you to the AC sine wave and how it was produced. The benefit of AC is that it can be easily transformed from one voltage level to another. This chapter will introduce you to the effects AC has on a circuit that are not experienced with DC systems. Specifically, the effect that induction has on a wire that is coiled up creates a significant difference. As we have already learned and reviewed, the three factors needed to induce a voltage into a conductor are a conductor, a magnetic field, and relative motion. We will have all three of these factors affecting every coil of wire when AC is applied but not when there is a steady current such as with DC. This means the same wire reacts differently to AC than it does with DC. The effects of self-induction are explained!

OBJECTIVES

After completing this chapter, you should be able to:
- Define inductance
- Determine when self-inductance is used and the effects
- Calculate inductive reactance under various conditions
- Determine the angle between resistance and inductive reactance
- Determine the impedance of a coil
- Apply Ohm's law equations to determine circuit quantities

INDUCTANCE

Inductance is the ability of a current-carrying conductor, coil, or circuit to induce a voltage into itself or adjacent circuits. Self-inductance is the ability to induce a voltage into itself. There is some self-inductance present in all AC circuits because of the continuously changing magnetic fields.

INDUCTORS

A coil of wire wrapped around a core is called an inductor. Its name **inductor** comes from the fact that a voltage can be *induced* into it by moving it through a magnetic field or by moving a magnetic field around it. This induced voltage produces a current. You also learned that passing current through a wire produces a magnetic field around it. Inductors have many characteristics that are present with AC current connected and are not present with DC current connected. The same inductor with 100 V DC may have 10 A DC total current. The exact same coil connected to 100 V AC may only have 1 A of total current at 60 Hz and even less if the 100 V AC is at 100 Hz. The AC effects on inductors and the way inductors affect the AC circuits are discussed next. The schematic symbol for the inductor is shown in Figure 3–1.

FIGURE 3–1   Inductor symbols with different core types.

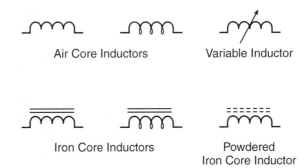

Air Core Inductors   Variable Inductor

Iron Core Inductors   Powdered
                            Iron Core Inductor

THEORY OF SELF-INDUCTION

As you know, whenever there is a current flow in a conductor, there is a magnetic field produced. When AC is used, the current and therefore the magnetic field is constantly increasing, decreasing, or reversing. The effect is that the magnetic field is always moving. As you recall, the factors needed for generation of an induced voltage are (1) a magnetic field, (2) a conductor, and (3) relative motion between the conductor and the magnetic field. The effects of generation are very apparent as we make a single conductor into a coil so that the conductors are on top of each other in close proximity (Figure 3–2). The result of this close physical location is that the magnetic field developed in the wire as it is expanding and contracting following the current sine wave will actually expand or "cut through" the neighboring conductor. Likewise, as the expanded magnetic field contracts, it cuts through the neighboring conductor

Inductance

The property of an electric circuit displayed when a varying current induces an electromotive force in that circuit or in a neighboring circuit. A circuit has inductance when magnetic induction occurs.

Inductor

Inductor is the name given to an electrical circuit component that exhibits the properties of inductance. If a coil of wire creates self-induction creating a counter electromotive force (CEMF), then it is referred to as an inductor.

FieldNote!

Many of the terms used in the electrical industry are synonymous. It just depends on where you learned the theory and what part of the country you're from. The term *inductor* is used to describe a coil or a particular piece of equipment that exhibits inductive characteristics. Variable inductors may have moveable cores. Inductors may be deliberately designed to choke off high current as in the case of power supplies, in which case the same component is referred to as a *choke*. Inductors are important components in AC electrical systems.

FIGURE 3-2

A single conductor coiled to make an inductor.

Self-inductance
The property of an electrical component (such as a coil of wire) to induce a voltage into itself as the current through the component changes.

again. As it magnetically moves through the conductor it will generate a voltage in the conductor. Just as in a generator, the small voltages in each segment add up to a larger voltage total in the length of the wire that is coiled up. This effect in a coil of wire with AC current flowing is called **self-inductance**.

Consider the coil shown in Figure 3–3. The magnetic field created by the changing current through the conductor expands and collapses with the current. This changing magnetic field cuts through the turns of the coil, inducing a voltage in the coil that opposes the change in current. Electrical loads, such as motors and transformers, contain these coils.

FIGURE 3-3   Cutaways of a coil with representative magnetic lines of force interacting.

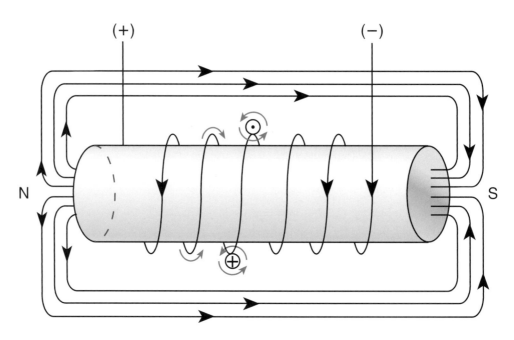

The total effect of all this can be summarized in three statements:
1. As the current increases, the magnetic field that it produced builds (increases).
2. As the current decreases, the magnetic field that it produced collapses (decreases).
3. When the current changes direction, the magnetic field it produced changes magnetic polarity.

Table 3–1 illustrates the entire process and introduces some very important considerations. This table shows that as the magnetic field collapses, it creates a voltage that opposes the original change in voltage. If the current in the coil decreases, the collapsing magnetic field cuts through the coil conductors and induces a voltage that tries to prevent the decrease. If the current in the coil increases, the expanding magnetic field cuts through the coil conductors and induces a voltage that tries to prevent the increase. The magnitude of the opposing voltage created will be determined by how fast the current is changing. Remember that the voltage induced is a function of how many lines of flux are being cut each second. Thus, this overall effect is one in which the increasing or collapsing magnetic field induces a voltage that is 180° out of phase with the applied voltage.

TABLE 3–1   Magnetic Field Action in an Inductor and Induced Electromotive Force (EMF)

Applied Voltage	Current Flow	Magnetic Field Created by Current	Voltage Induced by Magnetic Field Change
Constant (DC)	Constant	Constant	Zero
Decreases	Starts to decrease	Decreases	Opposes change
Increases	Starts to increase	Increases	Opposes change

For example, assume an inductor with no resistance is connected to a 120-V AC supply. Further assume a point in time when the collapsing (or increasing) magnetic field is inducing a voltage of 108 V into the coil. Since an equal amount of applied voltage must be used to overcome the induced voltage, there will only be 12 V to push the current through the conductor. Figure 3–4 shows the 180° phase relationship between the applied voltage and the induced voltage as well as the vector calculations for the resultant voltage.

FIGURE 3–4   Vector representation of EMF opposing CEMF with the resultant voltage.

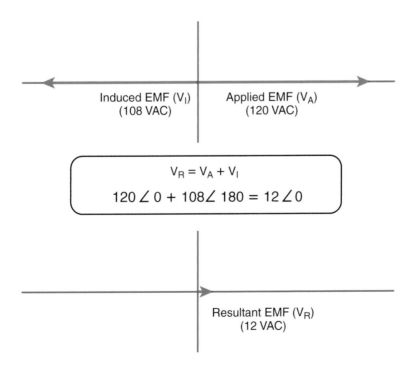

Induced EMF (V$_I$)
(108 VAC)

Applied EMF (V$_A$)
(120 VAC)

$$V_R = V_A + V_I$$

$$120 \angle 0 + 108 \angle 180 = 12 \angle 0$$

Resultant EMF (V$_R$)
(12 VAC)

CEMF (Counter EMF)
The voltage induced in an inductor by the changing magnetic field. The voltage induced is "counter to" the voltage, or in opposition to, the voltage that produced the original magnetic field.

Since the induced EMF appears only during a change in the current flow, the magnitude of the self-induced EMF depends on the rate of change of the current and the amount of the inductance. The induced EMF is also called the counter EMF and is abbreviated **CEMF.** As indicated in the formula in Figure 3–4, the 120 volts of applied voltage is the reference point at an angle ($\angle$) of 0°. Add to it a CEMF of 108 volts at 180°, and the opposing vectors cancel except for the remaining 12 volts resultant at 0°.

LENZ'S LAW

Heinrich Lenz, a German physicist, discovered that a magnetic field that induces voltage and a magnetic field that is produced by the induced current oppose each other. This is the basic principle used to determine the direction of an induced voltage. Lenz's law states that an induced voltage or current opposes the motion that created the induced voltage or current. The induced voltage and related current is 180° out of phase from the applied voltage and resulting current.

When the current through an inductor changes the magnetic field around the inductor will change. This change in magnitude causes the field to move (expand or collapse) relative to the coil. The relative motion will induce a voltage that opposes the change in current that produced the change in the field. Figure 3–5 shows the phase relationship between the applied voltage and the induced voltage. The basic rule is that inductors always oppose a change in AC current.

FIGURE 3–5   Graphical representation of EMF and smaller CEMF 180° out of phase.

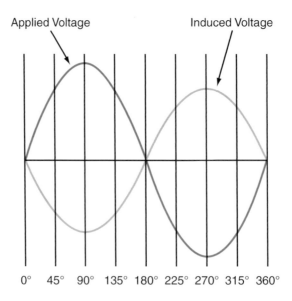

It is important to understand the concept of applied voltage (EMF) and induced voltage (counter EMF) and their respective currents as they relate to the magnetic field surrounding a conductor. Figure 3–6 may help explain that relationship. The following explanations reference the sine wave shown in Figure 3–6 and the numbered points along that waveform.

FIGURE 3–6 The applied current waveform and resultant magnetic flux directions.

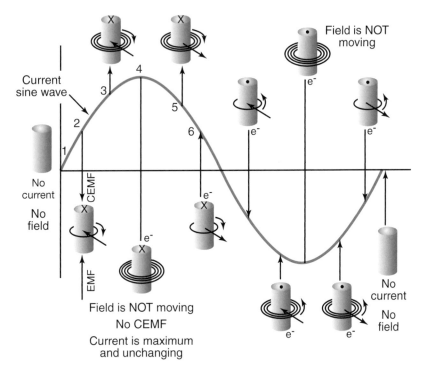

1. With no current flowing, there is no field.

2. As the current flow begins to increase, the field begins to expand, cutting the conductor in the direction shown (moving outward).

3. As the current flow continues to increase, the field continues to expand.

4. The current is at its maximum and is momentarily unchanging. The field is also not changing (not moving) because of the stable state of the current flow.

5. After having been at the maximum, the current flow begins to decrease (change), and the magnetic field begins to react by collapsing inward (changing the direction in which it is moving relative to the conductor). The field does not change polarity, just strength.

6. The current continues to decrease, and the field continues to decrease relative to the current flow.

7. The current will change direction and is increasing but this time in a negative direction. Since the current is increasing in the opposite direction, the magnetic field expands outward but with reverse polarity (polarity opposite to what was found during the positive portion of the sine wave).

8. As the current flow continues to expand, the field also continues to expand.

9. The field is at its maximum as a result of the current being at its maximum negative value. The magnetic field is neither expanding nor decreasing at this point.

10. After having been at the maximum negative value, the current value begins to decrease (become less negative), and the field will also continue to decrease (collapse inward, changing the direction in which it is moving relative to the conductor); however, the polarity will not change.

11. As the current continues to decrease, the field will also continue to decrease.

12. With no current flow, there is no field.

HENRY

The term *inductance* is used to describe the property of a coil to produce self-induction. Therefore, a coil has the property of inductance if it will react to an AC current in a specific way. We can predict how the coil will react to a current based on its physical properties. A formula allows us to produce a coil that has a certain amount of inductance which is measured in **henries.** It is named after the American scientist Joseph Henry. Based on the coiled wire around a circular core, the formula for a henry, symbol L, is:

$$\text{English units: } L = \frac{3.19 \, N^2 \times A \times \mu}{10^8 l}$$

or:

$$\text{Metric units: } L = \frac{0.4 \times \pi \times N^2 \times A \times \mu}{l}$$

Note: These formulas are for circular cores only. Other core shapes use similar but different equations.

L in henrys

$0.4 \, \pi$ is a constant for a circular core or 3.19 for English units

N is number of loops or "turns" of the conductor

A is the area of the core in square meters or square inches for English units

μ (lowercase Greek letter mu, pronounced *mew*) is the magnetic **permeability** of the core material

l is the length of the core in meters for the metric formula or inches in the English units formula

PHYSICAL FACTORS AFFECTING INDUCTANCE

There are a number of factors that determine the amount of inductance in a coil. The following sections describe those factors.

3.1 Number of Turns of Wire

The self-inductance of a wire or arrangement of wires is directly related to the number of times the wire is coiled. To understand why this is, start by looking at Figure 3–7. In the figure, a conductor has been looped into one coil. Using the left hand rule for the magnetic field, you can see that the fields from each side of the coil add to the fields set up by their opposite sides. Therefore, by creating a single coil, the number of flux lines and hence the inductance have been increased.

Figure 3–8 shows what happens when more turns are added. The physical arrangement causes each individual coil's field to add to those of the other coils. Clearly, the total number of flux lines and the inductance are directly related to the number of turns of wire.

3.2 Spacing between the Turns

Notice in Figure 3–8 that not all of each coil's flux lines link all the other coils. Some of the flux lines from one coil may link with only one other coil,

Henry

Unit of measure for inductance. A coil has an inductance of 1 henry when a current change of 1 ampere per second causes an induced voltage of 1 volt. The henry is a value based on the physical characteristics of the inductor (coil). The unit of measure is the henry, abbreviated as H. The symbol used in formulas is L.

Permeability

The ability of a material to concentrate or focus magnetic flux lines. Absolute permeability is measured in henries per meter. The permeability of a vacuum is $\frac{1.26 \times 10^{-6} \, H}{m}$. The symbol used for magnetic permeability in a formula is the lowercase Greek letter mu (μ).

FIGURE 3-7 Magnetic flux around a single conductor and magnetic effect of a coiled conductor.

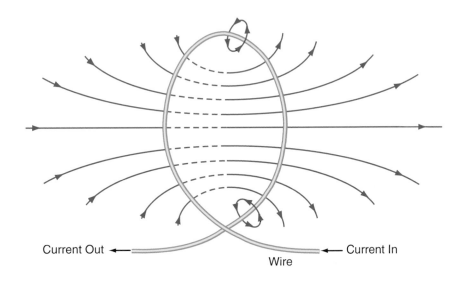

Current Out ◄─────────   ◄── Current In

Wire

FIGURE 3-8 Accumulative effect of multiple magnetic fields aiding in a coiled conductor.

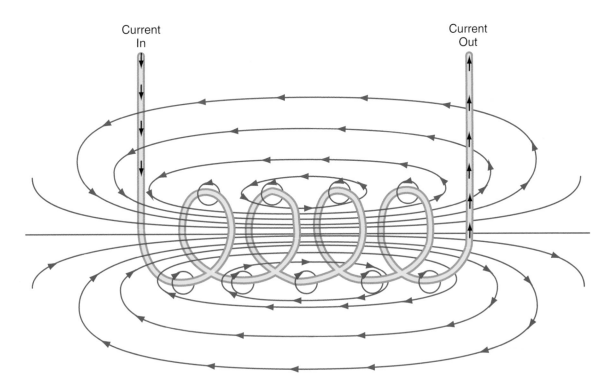

Current In

Current Out

two other coils, or perhaps none at all. The flux lines that do not link all the other coils do not contribute to total inductance and are called **leakage flux.** One way to reduce the amount of leakage flux is to wind each individual coil closer to the others.

Leakage flux
Flux lines that do not link properly in an inductor. Some flux lines are lost to the surrounding space and do not complete the magnetic path. Leakage flux reduces the overall magnetic field and the resulting inductance.

3.3 Cross-Sectional Area of the Core

The inductance of a coil is also related to the density, or concentration, of the flux lines. To understand this, think of the magnetic field of the Earth. The intensity of the Earth's magnetic field is immense, yet it does not "snatch the watch right off of your wrist." The reason is that the Earth's lines of flux are spread across a vast area. The total strength of any magnetic field is a function of both the number of lines and how dense they are.

3.4 Permeability of the Core Material

TABLE 3–2   Magnetic Properties of Ferromagnetic Materials

Material	Relative Permeability
Iron, 99.8% pure	150
Iron, 99.95% pure	10,000
78 Permalloy	8,000
Superpermalloy	100,000
Cobalt, 99% pure	70
Nickel, 99% pure	110
Steel, 0.9% C	50
Alnico 5	4

Magnetic permeability is the ability of a material to allow or conduct magnetic lines of flux. Iron has a permeability that is on the order of 1000 times greater than air. This means that if an iron core is inserted into a coil, the lines will be concentrated, and the resulting magnetic field and inductance will be increased. Table 3–2 lists the initial permeability of some ferromagnetic materials.

When all the effects of these various characteristics are evaluated and measured, the formula for henries can be used for a coil with a circular core:

- For other shaped cores (oval, square, or triangular), different equations apply.
- Inductance has no effect on a DC circuit as long as the current is constant. It does have an effect if the supply is turned off and on or when the DC current changes in value.

The amount of voltage induced into the neighboring conductor depends on the strength of the magnetic field and, as importantly, how fast the magnetic field moves. Also the direction of the induced voltage or EMF is of utmost importance. As you'll notice in the formula for CEMF below, the coil value in henries is multiplied by the change in current (ΔI—uppercase Greek letter delta, "I"—indicates change in current) divided by the change in time (Δt— uppercase Greek letter delta, "t"—indicates change in time). The negative sign in front of L indicates that the voltage is a negative, or counter, voltage. The more the current changes in amplitude, the more CEMF is generated. The faster the current changes the more CEMF is produced. Lenz's law paraphrased states: The direction of the induced EMF is always opposite to the voltage that created the original current in the coil. The opposite induced voltage is referred to as the counter (CEMF). This is sometimes referred to as the "back" EMF which has the exact same meaning.

The formula for determining how much CEMF is produced is:

$$V = \frac{-\text{L} \times \Delta I}{\Delta t}$$

where V is amount of voltage produced—actually CEMF
L is the henries of the coil
ΔI is the change in the current
Δt is the change in time

The amount of voltage is directly proportional to the changing amount of AC current in the conductor and inversely proportional to the change in the time that occurs. The rate of change of amps per second of the changing current will determine the amount of CEMF. The faster the frequency of the current waveform, the more CEMF is produced.

TechTip!

When the lights go down in a theater the voltage to the lighting is reduced through a dimmer. This used to be accomplished with an inductor. By changing the amount of iron in the core, by pushing the core in or retracting it so the core became more air and less iron, the inductance of the coil would change. If the core was removed, the inductance changed because more flux lines leaked out and did not cut neighboring conductors.

(continued)

Example

Find the henry value of an inductor with the following dimensions: The number of turns on the coil of wire is 100, the area of the core in the middle of the coil is 1 square inch, the permeability of the iron core is 150, and the length of the core is 2 inches.

Solution:

$$L = \frac{3.19\ N^2\ A\mu}{10^8 l}$$

$$L = \frac{3.19 \times 10,000 \times 1 \times 150}{100,000,000 \times 2}$$

$$L = .0239 \text{ henries or } 23.9 \text{ millihenries.}$$

Use the same coil of wire but insert a different core material with twice as much permeability and note that the henry value increases.
Change the core material to air, or no metal material, and the permeability goes to 1 and the inductance is $\frac{1}{150}$ of the original, or $\frac{.0239}{150} = .0001595$ henries.

Example

If a 1 L coil has a changing current of 1 A in 1 second, the CEMF is a −1 V as proven by the formula:

$$\text{CEMF} = \frac{-L \times \Delta I}{\Delta t}$$

$$\text{CEMF} = -1 \times \left(\frac{1\text{A}}{1 \text{ sec}}\right) = -1 \text{ V}$$

If the same henry coil has a changing current of 2 A in 1 second, then the CEMF is −2 V. As proven by the formula:

$$\text{CEMF} = -1 \times \left(\frac{2\text{A}}{1 \text{ sec}}\right) = -2 \text{ V}$$

If the same coil has a 1 A change in 2 seconds, the CEMF is −½ V. As proven by the formula:

$$\text{CEMF} = -1 \times \left(\frac{1\text{A}}{2 \text{ sec}}\right) = -\frac{1}{2} \text{ V}$$

INDUCTIVE REACTANCE

Because this CEMF is in opposition to the original (applied) voltage, it tends to oppose the current that is flowing in the coil, making it harder to get current to flow in the original desired direction. Because there is an opposition to current flow it was originally characterized as a resistance and was measured in ohms. *Be aware* that even though this is still measured

TechTip! *(cont'd)*

Therefore, the self-induction would change because of the core position and the amount of CEMF and the resultant current flow would be affected. As electronics became more prevalent, the old theater dimmers were mostly replaced. The concept of changing the henry value by adjusting the core is still used in various applications.

Interesting Note: During the performance, theater lights are not completely powered off; even when the lights are called to be off, the dimmer system keeps a very small amount of voltage and current running through them. This is so small that no light is produced. This is known as "pre-heat" and is done to keep the bulbs from cooling down before being turned on again and again. This saves lamp life.

TechTip!

One of the concepts to keep in mind is the fact that a coil does not change its henry value with a change of frequency or a change in the current value. A 100 millihenry coil is still a 100 millihenry inductor whether you apply DC or 60 Hz AC or 100 Hz AC. The way it reacts to the circuit will change but not the inductive value, unless you change the physical characteristics.

in ohms it is not really a resistance but rather a reaction to the effects of the CEMF. This reaction due to induction in an inductor is known as inductive reactance. The inductive reactance is measured in ohms as expressed in the formula:

$$X_L = 2 \times \pi \times f \times L$$

where X_L is inductive reactance measured in ohms
$2 \times \pi$ is a constant for sine wave fluctuations
F is the frequency of the AC measured in hertz
L is the inductance of the coil measured in henries

TechTip!

In order to make it easier to use your calculator you might take a shortcut. Whenever you are using the quantity (2π F) and the standard frequency is 60 Hz as it is in the United States, a rounded figure of 377 will usually suffice. Unless you are looking for precise values, the factor of 377 instead of 2π F will get you very close to the correct values.

Example

Examples of $X_L = 2 \times \pi \times f \times L$

Find the inductive reactance of a coil if the .5 henry coil is connected to 120 V AC at 60 Hz:

$$X_L = 2 \times \pi \times 60 \text{ Hz} \times .5 \text{ H} = 188 \text{ } \Omega$$

Connect the exact same coil to a 120 V, 120 Hz power supply:

$$X_L = 2 \times \pi \times 120 \text{ Hz} \times .5 \text{ H} = 376 \text{ } \Omega$$

Find the frequency of a 1 H coil connected to AC with an inductive reactance of 150 Ω:

$$X_L = 2 \times \pi \times f \times L$$

or

$$\frac{X_L}{2 \times \pi \times L} = f$$

$$\frac{150 \text{ } \Omega}{2 \times \pi \times 1} = 23.87 \text{ Hz}$$

Suppose you calculate the inductive reactance of a coil as .5K Ω. The frequency is 60 Hz as observed on an oscilloscope. Calculate the inductance of the inductor:

$$X_L = 2 \times \pi \times f \times L$$

$$\text{or L} = \frac{X_L}{2 \times \pi \times f}$$

$$L = \frac{500 \text{ } \Omega}{2 \times \pi \times 60} = 1.32 \text{ H}$$

LAGGING CURRENT

The reaction of the CEMF on the circuit is very apparent. For example, if you measure the actual *resistance* of a coil of wire with an ohmmeter, you might measure 6 ohms. If we supply 120 V AC to that coil we might assume there would be 20 A, according to Ohm's law $\left(\dfrac{120 \text{ V}}{6 \ \Omega} = 20 \text{ A} \right)$. However, the actual current reading is much less than that, only 12 A. There is more opposition to the AC current than we can measure with an ohmmeter. The "other" opposition is the inductive reactance. We have seen how to calculate this value but we cannot measure it directly. The inductive reactance actually creates a condition that causes the current to lag behind the voltage by 90 electrical degrees. The explanation for this is as follows.

As the current in the circuit is constantly varying from zero to maximum and back to zero, the magnetic flux around the conductor is also varying from zero to maximum and back to zero. As the flux lines reach their maximum value, corresponding to the line current, the magnetic lines are strongest and will create the greatest CEMF in the same conductor. The greatest CEMF creates the most opposition, and in a perfect circuit would stop the flow of current. Since the voltage and the original flux are moving in unison in the conductor, the maximum flux will produce the maximum opposition, therefore causing the least amount of current. There is more applied voltage than there is CEMF so the applied current continues to flow. Sometimes this difference is referred to as the differential value. See Figure 3–6 to review.

FIGURE 3-9   Current "lags" the voltage in a perfect inductor.

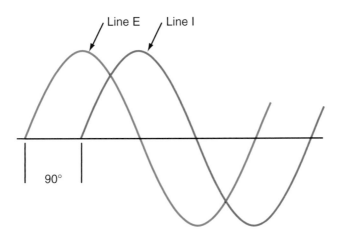

By referring to Figure 3–9 you can see that the actual current lags the line voltage by 90°. In a perfect inductor the current lags behind the applied voltage by the 90 electrical degrees indicated. Since this lagging effect is the result of the CEMF, caused by inductive reactance, the inductive reactance is shown as lagging the actual wire resistance of the circuit by the same 90° in a vector diagram and an opposition triangle. We now have a total opposition right triangle known as the impedance triangle (Figure 3–10).

Impedance
The vector sum of the oppositions found in some AC circuits. The vectors may include inductive reactance, capacitive reactance, and resistance. The vector addition of these components will result in the total opposition to the AC current flow. Impedance is measured in ohms and is represented by the letter Z in formulas.

FIGURE 3-10 The impedance triangle has components of resistance and inductive reactance, and the resultant hypotenuse is impedance. All are measured in ohms.

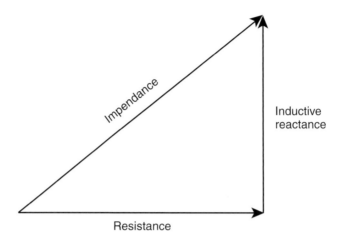

IMPEDANCE

To understand the calculations needed to predict the behavior of inductors in an AC circuit you must understand the mathematics of right triangles and associated elementary understanding of trigonometry. As mentioned before, the X_L of a coil is measured in ohms. The resistance of the wire in the coil is also measured in ohms. The total opposition is measured in ohms as well. The total opposition to the AC current is called the **impedance** and is noted in calculations as Z. The impedance is not just an arithmetic sum of the two oppositions. Instead, it is the vector sum of resistance (R) and inductive reactance (X_L) (Figure 3–11).

FIGURE 3-11 Z is the vector sum of R and X_L.

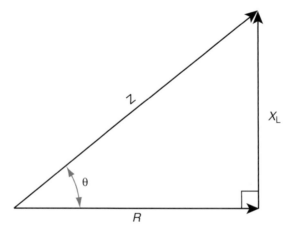

As mentioned earlier, the inductive reactance X_L, or CEMF is 180° out of phase with the applied voltage and causes the current to lag behind the applied line voltage by 90° (Figure 3–12). As you know, the actual wire resistance opposes current in the circuit and is considered "in phase" or is drawn on the 0° reference line as "R". The CEMF, which is a result of X_L, would cause the current to lag the voltage by 90°. This opposition is drawn at 90° to the horizontal and labeled as X_L as shown in Figure 3–13. Using trigonometry and the right triangle, we can calculate the actual angle of lagging current. It is between 0° and 90° as represented by the angle theta. The opposition to current that the total circuit exhibits is calculated by how it impedes the current. The total opposition is referred to as impedance.

FIGURE 3-12 The result of inductance would be to cause the current sine wave to LAG the voltage sine wave by 90°.

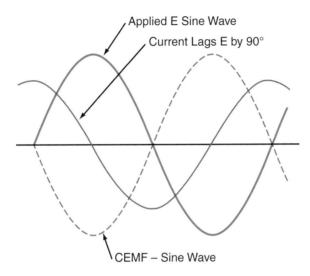

Applied E Sine Wave

Current Lags E by 90°

CEMF – Sine Wave

FIGURE 3-13 The AC oppositions forms a right triangle with R, X_L, and Z and angle Theta (θ).

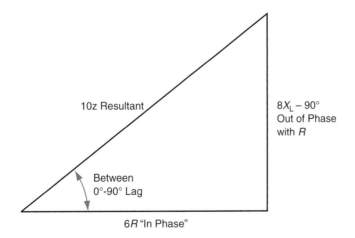

10z Resultant

$8X_L – 90°$
Out of Phase
with R

Between
0°-90° Lag

6R "In Phase"

From the original example of the coil that really only had 12 amps of AC current instead of the 20 amps expected, you can see that the hypotenuse of our triangle is 10 ohms of Z when the resistance is 6 ohms and the X_L is 8 ohms. The additional information that we can determine is that the current is no longer "in phase" with the voltage nor is it "90° out of phase" with the voltage but is somewhere between the two extremes. The actual value that the current lags behind the voltage in the circuit is found by using trigonometry to find angle theta in the impedance triangle in Figure 3–14. The number of degrees the current follows the sine wave voltage is the same as the angle between the R and the Z in the impedance triangle. See Appendix A following sections on Vectors, Scalars, Trigonometry, and Vector Addition.

FIGURE 3–14   Angle theta (θ) is the angle of lag between the voltage and current in the sine wave.

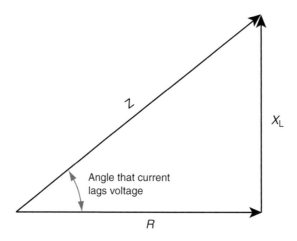

To help place all the above in context, the resistance causes part of the opposition and the resultant vector at 0° and can be measured with an ohmmeter. Likewise the X_L causes part of the opposition and the vector is drawn at 90° (matching the X_L) and is calculated by the formula for inductive reactance. The total current of the circuit is calculated by: $I_T = \dfrac{E_T}{Z_T}$ (Figure 3–15).

FIGURE 3–15   An AC circuit with an inductor will have current calculated by using the voltage and the circuit impedance.

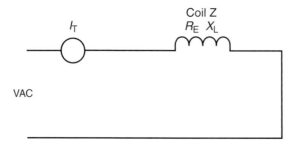

The I_T can be measured as the current of the circuit with an ammeter and is really the current due to the Z of the circuit.

Example

Example of Z in other applications:
A coil has a calculated value of 120 Ω of X_L. The same coil has a measured resistance of 50 Ω. Find the impedance of the coil.

$$Z = \sqrt{R^2 + X_L^2} \quad \text{(Pythagorean theorem)}$$

$$Z = \sqrt{50^2 + 120^2}$$

$$Z = \sqrt{2500 + 14{,}400} = 130 \ \Omega$$

Connecting 120 V AC to the coil would result in an ammeter reading of .92 A or $\dfrac{120 \text{ V applied}}{130 \ \Omega \text{ of impedance}}$. Measuring across the coil with a voltmeter would read 120 V. To verify the impedance use $\dfrac{120 \text{ V}}{.92\text{A}} = 130 \ \Omega$. You cannot measure 130 Ω with a standard ohmmeter. We will investigate the phase angle in more detail later but the actual angle of lagging current is not 90°, as we vectored the inductive reactance compared to the resistance. Instead the angle that the current lags the voltage through the coil is the same angle the impedance vector makes in relation to the resistance vector. In the above problem, the angle θ (lowercase Greek theta) is found by trigonometry. Use the arc tan function to find the arc that has the tangent of $\dfrac{120}{50}$ or 2.4. That arc is 67.3°. This means the current that flows through the coil is 67.3 electrical degrees behind (lagging) the applied voltage (Figure 3–16). Since all the sides of the impedance triangle are known, you can use any trig function to obtain the angle theata.

Do you think the current would double if the frequency were to be reduced to 30 Hz?

Even though the inductive reactance would go to half of the original value, the value does not change at the same rate because it is the hypotenuse of the triangle or the vector sum of R and X_L.

FIGURE 3–16

An impedance triangle with the current lagging the voltage by the angle θ of 67.3°.

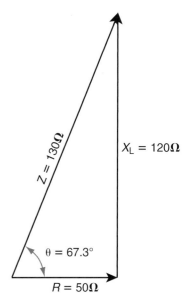

SUMMARY

This chapter took the AC waveform and applied it to a conductor that was coiled. The coil may be loosely coiled or tightly coiled on a metal core. The AC magnetic field caused the coil to create self-induction within the coil causing a CEMF to be induced. The coil, or inductor, produces the back voltage which acts like an opposition to the applied current. The opposition is measured in ohms of inductive reactance, or X_L. The inductive reactance is vectored in drawings of the vectors as a 90° opposition and the ohms of resistance as shown at a 0° vector. The resultant of the vector addition is the total opposition to the current and is referred to as the impedance of the inductor which is measured in ohms. The formula symbol is the letter Z. The behavior of the CEMF is expressed in Lenz's law. This principle of X_L, Z, and the effects of induction is critical to understanding how electric equipment powered by AC really works. The idea that an angle developed by the reactance and impedance of the coils is essential to understanding AC theory.

REVIEW QUESTIONS

1. The unit of measure for inductance is the _____. It is determined by the formula _____.

2. Explain what is meant by inductance.

3. If AC is applied to an inductor, then DC is applied to an inductor. Will the inductance increase, decrease, or stay the same?

4. Why is CEMF produced in a coil of wire with AC applied?

5. CEMF is a(n) _____ to applied current and is measured in _____.

6. Determine the CEMF produced in a 2-henry coil if the current is changing at a rate of .5 A every 2 seconds.

7. If you increase the frequency of the voltage on a coil, will the reactance increase, decrease, or stay the same?

8. Find the inductance in H of a coil that has 450 Ω of X_L when 120 V AC 60 Hz is applied.

9. An inductor has 500 turns of copper wire wound on a soft iron core 10 inches long and 3 inches in diameter. The permeability of the iron is 100. 100 V 60 Hz is applied. Find the inductance.

10. Basically, Lenz's law states: _____ _____.

11. What does impedance mean?

12. If a coil has a resistance of 30 Ω and a Z of 50 Ω, find the reactance at 60 Hz.

13. Using the coil in question 12, if we install a core of the same size but a higher permeability, will the impedance increase, decrease, or stay the same?

14. What type of instrument is used to measure X_L?

15. Why will the current in an inductor always lag the voltage?

Resistors and Inductors in Series

4

OVERVIEW

Inductors play an integral part in the performance of AC circuits. Usually the inductors are placed in a circuit to provide required functions. Occasionally the placement of an inductor is unintended and the results are unintentional. Either way an understanding of the interaction of inductors and their circuit characteristics is necessary to gain an understanding of AC fundamentals.

This chapter covers the way in which inductors react in a circuit when they are connected in series. Most of the formulas in this chapter depend on each inductor acting independently of the others in the circuit. The inductors are usually connected in the circuit so that the magnetic field from each one does not induce voltage and current into the others. If the inductors are in close proximity to each other the effects of aiding and opposing fields are discussed. The voltage that a coil induces into itself is caused by the characteristic of self-inductance. The effects of this self-induction require us to calculate many of the circuit characteristics since we cannot measure it directly. The inductance affects the opposition to current, the way the voltage is distributed, and even the way power is calculated and measured.

It takes the added variable of time for voltage and current to change in an AC circuit. The time it takes for current to build up to its maximum value in a coil is based on an exponential curve called the LR time constant curve. It has been established that the number of time constants for a particular inductor and resistor combination is always five time constants. The actual elapsed time in seconds is determined by the L and the R of the circuit.

OBJECTIVES

After completing this chapter, you should be able to:

- Solve for total inductance in a series circuit containing more than one inductor
- Find total inductive reactance in a series circuit containing more than one inductor
- Solve for unknowns in circuits containing more than one inductor in a series when the values of other variables are specified
- Solve for the total inductance of series inductors when the magnetic fields are aiding or opposing
- Determine all the calculated values for oppositions, voltages, and currents in a series RL circuit
- Determine the dissipated power, the apparent power, and the reactive power of a series RL circuit
- Compute the power factor, the phase angle, and components of the power triangle
- Determine the time required for one time constant and the time for a complete inductive current charge or discharge

INDUCTORS CONNECTED IN SERIES

When inductors are connected in series, the total inductance is the sum of the inductances of all the inductors. If the magnetic field of one inductor does not affect the other coil's magnetic field, then the effect on the circuit is similar to resistors connected in series. Because coils offer opposition to the current due to their wire resistance and their inductive reactance, the effect is equivalent to adding resistors in series. The opposition offered to an AC current will be expressed as the impedance (Z) of the coil. Before we see the total effects, we need to determine how the physical coil determines the total inductance of the circuit. The total henry value is calculated by the formula:

$$L_T = L_1 + L_2 + L_3 + \ldots L_N$$

where N is the total number of inductors in the circuit.

We now know that inductive reactance is represented as a vector drawn at 90° to the horizontal. This reactive opposition is dependant on the inductance of the coil and the frequency of the circuit. Each coil in a series circuit will have the same frequency applied. We can add all the inductive reactances (X_L) because they are all at the same angle (90°). See "Vector Addition" in Appendix A.

It is easy to show that the total inductive reactance in a series inductive circuit is also equal to the sum of the individual reactances. Start by remembering the formula for calculating the inductive reactance if the inductance and frequency are known:

$$X_L = 2\pi f L$$

then

$$L = \frac{X_L}{2\pi f}$$

Substituting into the above Equation yields:

$$\frac{X_{LT}}{2\pi f} = \frac{X_{L1}}{2\pi f} + \frac{X_{L2}}{2\pi f} + \frac{X_{L3}}{2\pi f} + \frac{X_{LN}}{2\pi f}$$

Multiplying the previous equation by $2\pi f$ gives:

$$X_{LT} = X_{L1} + X_{L2} + X_{L3} + \ldots X_{LN}$$

Example

The circuit in Figure 4–1 shows three inductors connected in series. L_1 has an inductance of 0.3 H, L_2 has an inductance of 0.5 H, and L_3 has an inductance 0.6 H. What is the total inductance of the circuit?

Solution:

$$L_T = L_1 + L_2 + L_3$$

$$L_T = 0.3 \text{ H} + 0.5 \text{ H} + 0.6 \text{ H} = 1.4 \text{ H}$$

$$L_T = 1.4 \text{ H}$$

FIGURE 4-1   **FIGURE 4-1**   Three inductors connected in series to a 70Hz AC source.

$$L_T = L_1 + L_2 + L_3$$

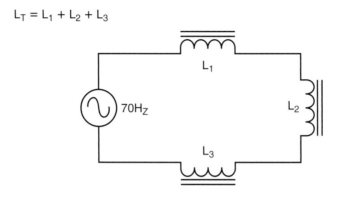

Example

The inductors in Figure 4–1 (L_1, L_2, and L_3) have inductive reactance of 132 Ω, 220 Ω, and 264 Ω. What is the total inductive reactance of the circuit?

Solution:

$$X_{LT} = X_{L1} + X_{L2} + X_{L3}$$

$$X_{LT} = 132\ \Omega + 220\ \Omega + 264\ \Omega = 616\ \Omega$$

Adding inductors in series with a circuit that already contains inductors will increase the total inductive reactance of the circuit. Also, remember from Chapter 3 that inductive reactance is measured in ohms (like resistance) and can be calculated when using Ohm's law as substituted for opposition.

Example

Prove that the listed frequency of 70 Hz in Figure 4–1 is true based on the example on page 65 and the top example on page 66.

Solution:
The inductive reactance and the inductance are related by frequency:

$$X_{LT} = 2\pi f L_T$$

Solving this equation for f and inserting the total inductance and inductive reactance for the previous examples yields:

$$f = \frac{X_{LT}}{2\pi(L_T)}$$

$$f = \frac{616}{2\pi(1.4)} = 70\ \text{Hz}$$

Note: f can also be proved by using one inductor (e.g. $L_1 = 0.3$ H & $X_{L1} = 132\ \Omega$).

Example

Assume the series circuit in Figure 4–1 has a 48-V power supply with a 400-Hz signal. Assuming no R in the coil, only X_L, what is the current?

Solution:

$$X_L = 2\pi fL$$

$$X_L = 2\pi \times 400 \text{ Hz} \times 1.4 \text{ H}$$

$$X_L = 3518.6 \ \Omega$$

Substituting X_L for R in Ohm's law gives:

$$I = \frac{E}{X_L}$$

$$I = \frac{48 \text{ V}}{3518.6 \ \Omega} = 13.6 \text{ mA}$$

Example

What is the value of the current if the frequency is dropped to 30 Hz?

Solution:

$$X_L = 2\pi fL$$

$$X_L = 2\pi \times 30 \text{ Hz} \times 1.4 \text{ H}$$

$$X_L = 263.9 \ \Omega$$

$$I = \frac{E}{X_L}$$

$$I = \frac{48 \text{ V}}{263.9 \ \Omega} = 182 \text{ mA}$$

FieldNote!

When doing quick approximations for the impedance of a coil the ratio of the inductive reactance to the resistance is considered. If the resistance is less than $\frac{1}{10}$ of the X_L, then we disregard the R and claim that the total opposition of the coil is made up of reactance. This helps speed the process of determining the oppositions in the circuit and also helps us approximate the degrees of lag that the current lags behind the voltage in inductive circuits. Remember, this process is for approximations only.

INDUCTORS WITH AIDING OR OPPOSING FIELDS

When coils are placed in close proximity to each other the magnetic fields generated by one coil with power applied may affect other coils nearby. The mutual induction is the characteristic that allows us to use transformers. Look at Figure 4–2 to see how coils may be placed on common iron core to increase the flux linkage. This is just an example to show how the magnetic fields may interact if the magnetic field of coil 1 is linked to coil 2. The two coils are connected in series so you would expect the total L_t to be the sum of L_1 and L_2. However, if you use the calculation $L_t = L_1 + L_2$ to verify this fact, the answers won't match. You will find the total L_t is either more than, or less than, the expected L_t. You will find that if the two coils are aiding each other, the magnetic field becomes

FIGURE 4–2

Two coils placed so that the magnetic fields interact with each other.

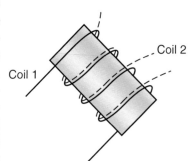

Coil 2

Coil 1

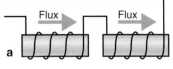

FIGURE 4-3

Mutual induction and the coefficient of coupling. (a) coils with fields aiding (b) coils with fields opposing and (c) coils at 90° to one another - neither aiding nor opposing fields.

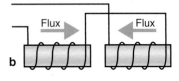

a

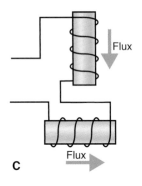

b

c

TechTip!

The two coils need to be parallel to each other in order for their fields to create a coefficient of coupling and affect the mutual inductance. By placing the coils perpendicular to each other, the magnetic fields do not interact and the coefficient of coupling or K is zero. Depending on whether you want the coils to interact or not may determine the physical placement of the coils.

stronger and the total L_t increases. If the fields are opposing each other, the magnetic field becomes weaker and the total L_t decreases. This is shown in the formula: $L_t = L_1 + L_2 + 2M$ if the fields are aiding and $L_t = L_1 + L_2 - 2M$ if the fields are opposing. M is the amount of mutual induction added or subtracted from the total sum of L_1 and L_2. The amount of mutual inductance (M) is calculated by the formula: $M = k\sqrt{L_1 \times L_2}$. The ($k$) in this formula is the decimal equivalent of the percentage of coefficient of coupling. In other words, what percentage of the magnetic fields of the coils are affecting each other? If the coils are on the same core and the fields interact 100%, then the $k = 1$. If there is no interaction because the coils are far apart, then the percentage of interaction is 0 and the coefficient of coupling is 0 and there is no effect on L_t.

For example: If the henry value of $L_1 = 3$ H and the henry value of L_2 is 3 H then we might assume the L_t of these two coils connected in series is 6 H. If the two coils are placed close to each other and they have 50% coupling of the magnetic fields and they aid one another then:

$$L_t = L_1 + L_2 + 2M \text{ (in Henrys)}$$

$$M = k\sqrt{L_1 \times L_2} \text{ (in Henrys)}$$

$$M = .5\sqrt{3 \times 3} = .5\sqrt{9} = .5 \times 3 = 1.5 \text{ H}$$

$$L_t = 3 + 3 + 2(1.5) = 9 \text{ H}$$

keeping the physical spacing of the coils the same and reconnecting the L_2 coil connections so that the magnetic field is opposite to what it was originally. Now the magnetic fields oppose each other. The formula becomes:

$$L_t = L_1 + L_2 - 2M \text{ (in Henrys)}$$

$$M = k\sqrt{L_1 \times L_2} \text{ (in Henrys)}$$

$$M = .5\sqrt{3 \times 3} = .5\sqrt{9} = .5 \times 3 = 1.5 \text{ H}$$

$$M = .5 \times 3$$

$$M = 1.5 \text{ H}$$

$$L_t = 3 + 3 - 2(1.5) = 3 \text{ H}$$

Example

Find the coefficient of coupling if the two coils connected in series are 5 H each. Assuming negligible resistance due to the wire in the coils, we find that the total X_{LT} is 4524 Ω at 60 Hz.

Solution:

Given two 5-H coils connected in series the L_T should be 10 H. The X_{LT} should be $X_{LT} = 2\pi fL$ or $(377 \times 10) = 3770$ Ω. Since the actual X_L is 4524 Ω, we conclude there is an additive field causing more inductive reactance. We must use the formulas:

$$X_{LT} = 2\pi f\, L_T$$

$$L_T = \frac{X_{LT}}{2\pi f}$$

$$L_T = \frac{4525\ \Omega}{377} = 12\ H$$

$$L_t = L_1 + L_2 + 2M \qquad \text{(Note: "}+2M\text{" is used because } X_L \text{ is larger than expected.)}$$

$$12H = 5\ H + 5\ H + 2M$$

$$(12H - 10\ H) = 2M$$

$$2H = 2M$$

Therefore: $1\ H = M$, or the mutual induction amounts to 1 henry.

The coefficient of coupling is:

$$M = k\sqrt{L_1 \times L_2} \text{ (in henrys)}$$

$$k = \frac{M}{\sqrt{L_1 \times L_2}}$$

$$k = \frac{1}{\sqrt{5 \times 5}} = \frac{1}{5} = .2$$

Therefore, the coefficient of coupling is 20%.

IMPEDANCE OF AN INDUCTOR

A quick review is provided here of the impedance calculation of a coil using the R of the wire resistance and the X_L of the inductive reactance to produce impedance. Sometimes we neglect the R of the coil if it is so small compared to the reactance that the total opposition of the coil, or Z of the coil, is essentially the X_L. However, if the R is comparatively large or the X_L is relatively small, then they must be included in the calculation.

Example

Assume that the coil in Figure 4–4 has an inductive reactance of 100 Ω and that the resistor has a resistance of 150 Ω. What is the total impedance?

Solution:

$$Z = \sqrt{R^2 + X_L^2}$$

$$Z = \sqrt{150^2 + 100^2}$$

$$Z = \sqrt{22,500 + 10,000}$$

$$Z = \sqrt{32,500}$$

$$Z \approx 180\ \Omega$$

FIGURE 4–4

A representative circuit with just X_L and R in series.

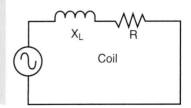

Q OF A COIL

The quality factor of a coil is referred to as the "Q" of the coil. The Q refers to the ratio of inductive reactance to resistance. A formula to find the Q is $Q = \dfrac{X_L}{R}$. The larger the ratio, or the more X_L there is compared to the R of the wire, the coil is considered higher quality. In practice for coils with a Q of 10 or greater, the R is ignored as being insignificant. If the Q is under 10 we should consider the R in our calculations.

Example

What is the Q of a .058H coil at 400 Hz and a resistance of 15 Ω?

$$X_L = 2\pi f L$$
$$X_L = 6.28 \times 400 \times .058$$
$$X_L = 145.77 \ \Omega$$

and

$$Q = \frac{X_L}{R}$$
$$Q = \frac{145.77}{15}$$
$$Q = 9.72$$

RESISTORS AND COILS IN SERIES

4.1 One Resistor and Coil

When adding resistive elements to a circuit that already has a coil, the oppositions to the current increase. A purely resistive circuit such as represented by an incandescent lamp will add R to the calculation. This "R" is drawn vectorially at a 0° angle to the horizontal as seen in Figure 4–5.

FIGURE 4–5   An incandescent lamp is a resistive element and is represented by a vector at 0°.

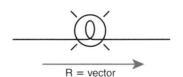

R = vector

This "R" is added to the "R" of the coil caused by the resistance of the wire in the coil. Remember you can actually measure this resistance with an ohmmeter. The X_L of the coil is also part of the opposition of the coil and is drawn at 90°. Now using vector addition, the "R of the lamp" plus the "R of the coil" add together arithmetically because they are at the same angle (0°). Vectorially add this to the X_L to get the triangle as shown in Figure 4–6. The hypotenuse of this new triangle is the impedance of the series circuit. The angle theta (θ) is now the angle of the hypotenuse to the horizontal. This angle now represents the actual angle of lag that the current lags behind the voltage in the circuit (Figure 4–7).

FIGURE 4–6   The R vector of the lamp is added to the R of the coil, and vectorially to the X_L.

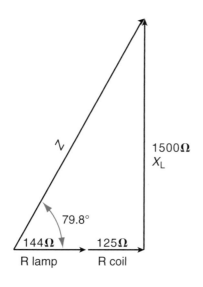

FIGURE 4–7   The Impedance right triangle contains the three components and the resultant total opposition at the circuit angle Theta.

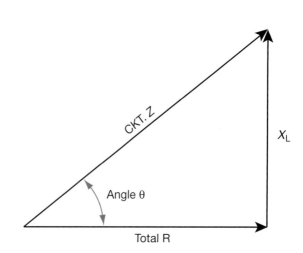

4.2 Resistors and a Coil

By adding another resistor (Figure 4–8), the circuit again changes. It is represented vectorially (Figure 4–9) with three resistive components and one X_L component. Notice that the triangle becomes flatter and the angle θ gets smaller. This means that the current is not lagging by as big an angle as before and the circuit behaves more like a resistive circuit.

FIGURE 4–8   The circuit diagram of a resistor a lamp and a coil in series.

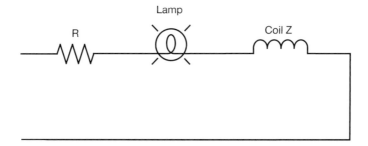

FieldNote!

Electric circuits are made of various combinations of three basic components; resistance, inductance, and capacitance. Two of these components are in many of the items we use everyday. The components are in series such as in electric motors or in other household and commercial equipment. Where the inductive reactance and resistance are both factors we need to consider how this affects the circuit. If the circuit is mostly resistive, as in a waffle iron or a toaster, the impedance is the same as the resistance and the impedance triangle appears flat or 0° of lagging current. If the load is a refrigeration motor, the impedance is made up of reactance and the impedance triangle has a larger angle θ, and the angle of lagging current moves more toward 90° lag.
Normally the lagging angle is not measured in the residential application, but that same angle is measured and may be charged for if it is too large in commercial accounts.

FIGURE 4-9   Vector diagram of the circuit represented in Figure 4-8.

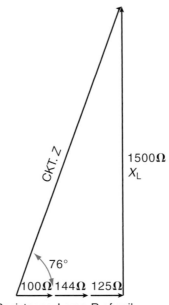

4.3 Two Coils in Series

Let us now examine two coils in series and see what happens as the total oppositions of each coil are added to the circuit. Coil one has 20 Ω of resistance and 30 Ω of inductive reactance (X_L). Coil two has 10 Ω of R and 40 Ω of X_L. Each coil has "R" and "X_L", as shown in Figure 4–10. Adding

FIGURE 4-10   Two coils with different characteristics connected in series.

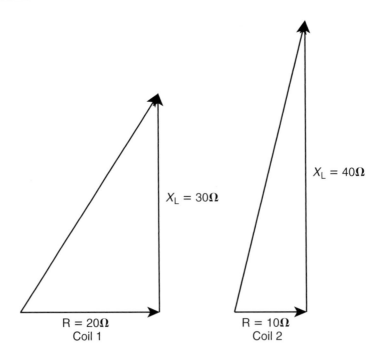

vectors together allows us to add R_{coil1} and R_{coil2} together and then add X_{coil1} and X_{coil2} together to get a triangle as shown in Figure 4–11. You cannot add the original "Zs" of the two separate triangles together because they are at different angles. The new large triangle now has a new Z_T which represents the new circuit total. This new circuit Z allows us to determine the actual current in the circuit by the formula: $I_T = \dfrac{E_T}{Z_T}$. We can also calculate the angle that the current lags the voltage in the circuit by the angle θ for the total circuit triangle. In this case the hypotenuse can be calculated by the Pythagorean theorem or by using trigonometry. The angle can be calculated by trig by using the arctan $\theta = \dfrac{\text{opposite side}}{\text{adjacent side}}$ or an angle of 66.8°.

4.4 Resistors and Inductors in Series

We can now add as many coils and resistive components as we like and use the same process of vector addition to construct a combined triangle with all the Rs and all the X_Ls in the proper angular positions. The resultant hypotenuse will be the total impedance of the circuit and the angle of the lagging current will be the final angle θ.

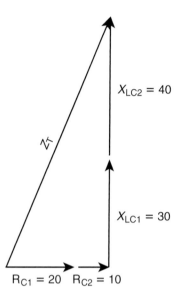

FIGURE 4–11

Impedance triangle representing R and X_L components of two coils.

$X_{LC2} = 40$

$X_{LC1} = 30$

$R_{C1} = 20$ $R_{C2} = 10$

Example

Three coils are connected in series. Each coil is slightly different in the amount of R and X_L. See Figure 4–12 for three coils. Coil 1 has 20 Ω of R and 20 Ω of X_L; coil 2 has 5 Ω, of R and 20 Ω of X_L; coil 3 has 25 Ω of R and 10 Ω of X_L. The vector addition for each of the coils is represented in Figure 4–13. The three R vectors are added to create a 50-Ω vector at 0°. The three inductive reactance vectors add up to 50 Ω at 90°. Therefore, the phase angle theta (θ) for the circuit is 45°. The current lags the voltage by 45° in this circuit. The total Z of the circuit is the hypotenuse of the triangle, using the Pythagorean theorem: $\sqrt{50^2 + 50^2} = 70.7\ \Omega$.

FIGURE 4-12   Three coils, each different, are connected in series.

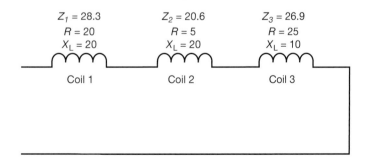

$Z_1 = 28.3$ $Z_2 = 20.6$ $Z_3 = 26.9$
$R = 20$ $R = 5$ $R = 25$
$X_L = 20$ $X_L = 20$ $X_L = 10$

Coil 1   Coil 2   Coil 3

FIGURE 4–13   Vector addition of three different coils as shown in Figure 4–12.

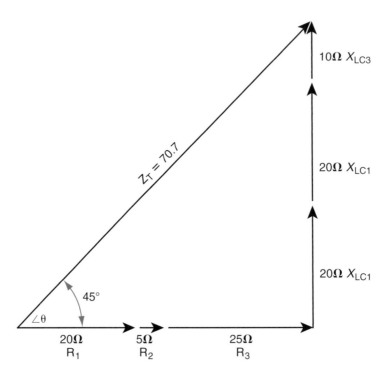

Example

Two coils and two resistors are connected in series as shown in Figure 4–14. In this example the two coils are identical and the two resistors are identical. Each resistor is 50 Ω and each coil has 25 Ω of R and 100 Ω of X_L. The vectors add up, as depicted in Figure 4–15, to yield an impedance for the circuit of 250 Ω. The angle of lag for the current in this circuit is 53.1°. Use trigonometry to find the angle. The angle between the adjacent side and the hypotenuse in Figure 4–15 can be calculated by the trigonometric function using the tangent, or more precisely "the angle or arc which has a tangent of," known as the arc-tan or on a calculator is shown as $\tan^{-1}$. The tangent of an angle is found by the formula $\dfrac{\text{opposite side}}{\text{adjacent side}}$. In this case the vertical side is the opposite side and the adjacent side is the horizontal side of the triangle, so the formula for finding the tangent of this angle would be $\dfrac{200}{150} = 1.333$. To find "the angle which has a tangent of" 1.333, use the arc-tan function. The angle which has a tangent of 1.333 is approximately 53°.

FIGURE 4-14 Two coils and two resistors connected in series.

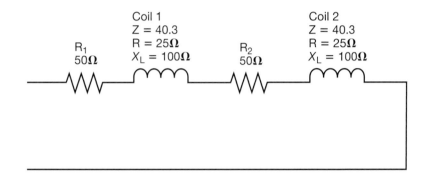

FIGURE 4-15 Vector addition of components in Figure 4–14 with the final resultant phase angle of 53° of lag.

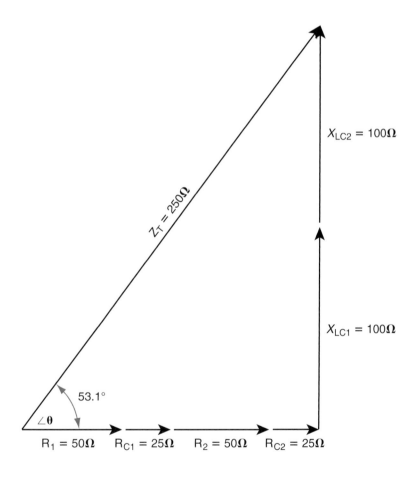

TOTAL Z, E, AND I AND COMPONENT Z, E, AND I IN A SERIES CIRCUIT

As you studied earlier, the impedance triangle changes as each new component is added. More resistance causes the triangle to become more flattened (smaller phase angle) and more X_L causes the triangle to become taller (larger phase angle). The phase angle changes as the right triangle shape changes. When the angle is small, the current is closer to a 0° angle and the current is more in phase with the voltage. As the angle becomes larger, the degree of lag between the applied voltage and the resulting current becomes larger and the current moves toward the 90° lag situation. The total Z is always calculated by adding the vectors or by using the Pythagorean theorem. The Z_T is used to calculate the total current and the angle θ determines the degrees that the current lags. This current is the same in every component of the series circuit. To calculate the voltage drop on any one component you will use the I_T and the total opposition of that particular component (Figure 4–16). This circuit has a 1 H coil and a 100 Ω resistor. The coil also has 50 Ω of resistance. The total circuit Z triangle is made up of resistance of 150 Ω and 377 Ω of inductive reactance when 60 Hz 120 V AC is applied.

The impedance of the circuit is 405.7 Ω. The current is $\dfrac{120}{405.7} = .295$ A and using the cosine function $Cos = \dfrac{R}{Z} = \dfrac{150}{405.7} = .36973$. The arc-cos, which has a cosine of .36973 or Cos^{-1} .36973, = angle of 68.3°.

FIGURE 4–16 Impedance triangle with a resistor and a coil and the Z total.

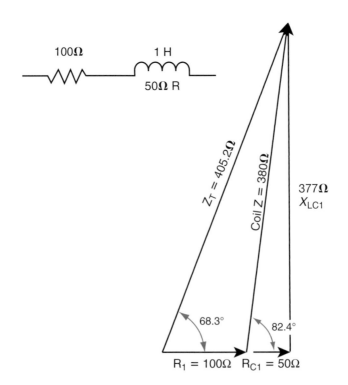

To calculate the voltage drop across just the resistor, use $I_T \times R_{resistor} = E_{resistor}$.

$$.295 \text{ A} \times 100 \text{ } \Omega = 29.5 \text{ V}$$

To calculate the voltage drop across the coil use $I_T \times Z_{coil} = E_{coil}$

$$.295 \text{ A} \times \text{approximately } 380 \text{ } \Omega \text{ of } Z = 112.1 \text{ V}$$

If you add the two voltage drops across the two components by simple arithmetic they do not equal the source voltage of 120 V AC. Instead you need to add the two voltages vectorially as illustrated in Figure 4–17. Now the voltage triangle resembles the impedance triangle and the resultant vector or the hypotenuse is the circuit total of approximately 120 V. The answer may not be exact as we have rounded some figures to ease in the calculation.

FIGURE 4-17 Adding component voltage vectors to get total circuit values.

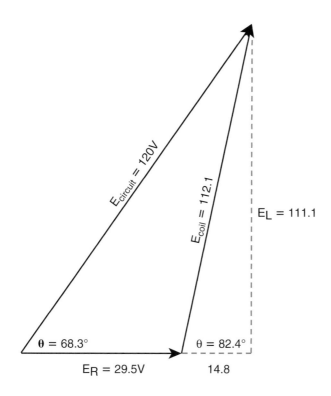

Step-by-step method to solve Figure 4–17:
1. The voltage across the resistor in the series circuit of Figure 4–15 is calculated by using Ohm's law when the current through the resistor (.295 A) and the resistance of the resistor (100 Ω) are known. The result is as follows:

$$100 \text{ } \Omega \times .295 \text{ A} = 29.5 \text{ V}$$

and is drawn on the horizontal line of the voltage triangle.
2. The voltage dropped by the coil uses the same circuit current multiplied by the coil impedance: 380 $\Omega \times$ .295 A = 112.1 V and is drawn at an angle of 82.4° to the horizontal.

3. One way to solve the final voltage triangle is to break the voltage of the coil into horizontal ($E_{R\ COIL}$) and vertical ($E_{L\ COIL}$) components.

4. The vertical component is found by using the sine of the 82.4° angle

$$\text{Sin } 82.4° = \frac{\text{Opposite}}{\text{Hypotenuse}}$$

Opposite = sin 82.4° × 112.1

Opposite = .991 × 112.1 = 111.1 V

5. The horizontal component of the voltage triangle is found by the cosine of 82.4° angle

$$\text{Cos } 82.4° = \frac{\text{Adjacent}}{\text{Hypotenuse}}$$

Adjacent = cos 82.4° × 112.1

Adjacent = .1322 × 112.1 = 14.8

6. Add all the horizontal components of the two components of 29.5 + 14.8 = 44.3 V

7. The final circuit voltage triangle has a horizontal base (E_{RT}) of 44.3 V and a vertical component of (E_{LT}) 111.1 V

8. The circuit voltage is the hypotenuse of the triangle
    Hypotenuse = $\sqrt{44.3^2 + 111.1^2} \approx 120$ V

When a circuit containing both resistance and inductance is connected to alternating current, the total current will lag the applied voltage at an angle between 0 and 90°. The exact amount of the phase angle difference is determined by the ratio of inductive reactance as compared to resistance. This is the opposite side of the impedance triangle compared to the adjacent side or $\frac{X_L}{R}$. The tangent of the angle θ is this decimal equivalent.

If a circuit contains a large amount of resistance and a small amount of inductance, the resistance and therefore the impedance will be nearly the same (Figure 4–18).

FIGURE 4–18   An impedance triangle of large circuit resistance and small inductive reactance yields a Z made up of mostly R.

$X_L = 1\Omega$
$R = 10\Omega$
$Z = 10.04\Omega$

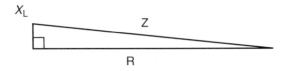

Also, if there is a large amount of inductance and a small amount of resistance, the impedance and inductance will be nearly the same (Figure 4–19) and is referred to as a ratio. There is a rule of thumb that if the ratio between the reactance and resistance is 10 or greater, the smaller value is considered negligible. You learned about the Q of a coil, which is defined as $Q = \dfrac{X_L}{R}$. When a coil has a large inductive reactance and a small resistance, it has a high Q.

POWER FACTOR, PHASE ANGLE, AND WATTS IN A SERIES CIRCUIT

We now have an impedance triangle and a voltage vector that are identical (congruent) shapes. To determine the power dissipated by each component as shown in Figure 4–16, refer to Figure 4–20. The power triangle will again resemble the impedance triangle.

The watts dissipated by the resistor will be the current times the voltage, $E \times I$, or use $I^2 R$.

$$.295 \text{ A} \times 29.5 \text{ V} = 8.7 \text{ W}$$

FIGURE 4–19

An impedance triangle of a large circuit reactance and a small resistance yields a Z made up of mostly inductive reactance.

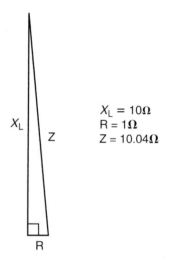

$X_L = 10\Omega$
$R = 1\Omega$
$Z = 10.04\Omega$

FIGURE 4–20 A power triangle is congruent to the voltage vector. The different components of power can be visualized.

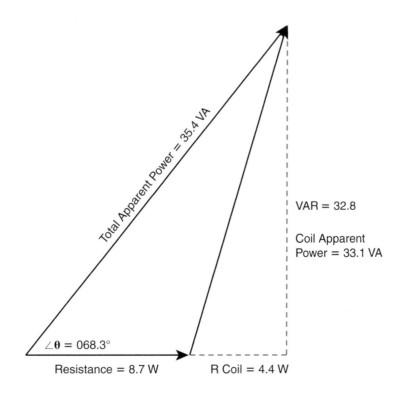

Total Apparent Power = 35.4 VA

VAR = 32.8

Coil Apparent Power = 33.1 VA

$\angle \theta = 068.3°$

Resistance = 8.7 W   R Coil = 4.4 W

FieldNote!

In geometry you can evaluate equivalent shapes of figures as the same shape with different actual dimensions. This concept is as if you were creating a scale model of one shape by changing the dimensions without changing the look of the original figure. In geometric shapes we say that the shapes are "congruent," meaning they have the same proportional sides, angular measurement, and shape, but may have different actual side dimensions.

The power dissipated in the coil is:

$$E_{\text{resistance.}} \times I_{\text{T}} = W_{\text{coil resistance.}}$$

$$.295 \text{ A} \times 50 \text{ } \Omega = 14.75 \text{ V}$$

$$14.75 \text{ V} \times .295 \text{ A} = 4.396 \text{ W}$$

The watts dissipated are $8.7 + 4.4 = 13.1$ W.

This is not the value you calculate when we multiply $E_{\text{T}} \times I_{\text{T}}$, as we did with DC circuits or purely resistive circuits. If you do use the totals of 120 V $\times$ .295 A = 35.4 volt-amperes (VA) you get a different side of the triangle, the hypotenuse. This answer of 35.4 VA is referred to as the **apparent power.**

Voltage and amperage are multiplied together to get volt-amperes (VA), which is the measure for apparent power. The actual power dissipated or consumed is less. The reason is that the voltage and the current waveforms are not synchronized, because they are 68.3° out of sync. Therefore we cannot just multiply current times voltage because the voltage and current waveforms don't reach peak or the zero point at the same time. Instead we have to consider **power factor.**

Power factor is a multiple (factor) applied to the apparent power to yield true power. It is also the cosine of the angle θ of the impedance triangle, the voltage triangle, or the power triangle. In our case the angle θ is 68.3° and the cosine of that angle is .369. We express power factor as a percentage, so cos. θ $\times$ 100 = % power factor. We have a circuit with a 36.9% power factor; this means 36.9% of the apparent power is converted to watts.

$$35.4 \text{ VA of apparent power} \times .369 \text{ power factor} \approx 13.1 \text{ W}$$

This 13.1 W is the watts dissipated by the resistor, and the resistance of the coil is referred to as **true power** and is measured in watts. The last step to solving the power triangle is to find the power developed across the inductive reactance of 377 Ω of X_{L}. This power is not consumed but is returned to the source. Remember that the X_{L} is caused by the magnetic field expanding and collapsing, creating counter electromotive force (CEMF). The movement of the magnetic field is energy and that energy is stored in the magnetic field. This power is referred to as **reactive power** and is measured in volts-amps-reactive (VARs). In this example we use the $I_{\text{T}} \times E_{\text{L}}$ or $I_{\text{T}}^2 \times X_{\text{L}}$ to get 32.8 VARs.

To verify the sides of the power triangle, add the true power of 13.1 W at 0° to 32.8 VARs at 90° to get a resultant of 35.3 VA, the apparent power of the circuit.

TYPES OF POWER

4.5 True Power

When an AC voltage is applied to a resistor, the current wave shape will be a copy of the voltage; that is, it will rise and fall at the same rate and in the same direction as the voltage and will reverse in polarity at the same time that the voltage reverses polarity. In this condition, the current is said to be in phase with the voltage (Figure 4–21).

This in-phase power component can be calculated by the $E \times I$ formula and is known as true power. It is also resistive power, or power dissipated as heat.

Apparent power

The amount of power that is delivered to an AC circuit. It is calculated by multiplying line voltage and line current. It is the amount of power apparently consumed, but is not the true power of a circuit with reactive components.

Power factor

A factor applied to the apparent power to yield true power. The power factor decimal is expressed as a percentage. It is also the ratio of watts divided by volt-amps expressed as a percentage of true watts compared to apparent volt-amps.

True power

The actual dissipated watts of a circuit. This is the form of energy expended by the conversion of electric energy into other forms of energy. It is the power measured in watts.

Reactive power

The form of power that is produced by the reactive components of a circuit, such as the inductive or capacitive reactance. It is energy stored in a magnetic field or an electrostatic field and is returned to the circuit as the fields diminish. Even though this is a form of energy for the circuit it is not consumed. It is measured in volts-amps reactive (VARs) because it is caused by the reactive current, 90° out of phase with the voltage.

FIGURE 4-21 AC voltage and current are "in phase" if they hit peaks and zero points simultaneously.

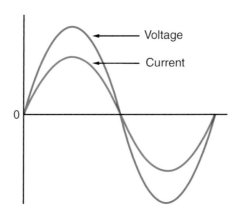

4.6 Reactive Power

In an inductive circuit, the current lags the voltage across the inductor by 90° (Figure 4–21) because the induced voltage across the inductor is 180° out of phase with the changes in source voltage. The current lags the voltage across the inductor by 90° (Figure 4–22) because the induced voltage across the inductor is 180° out of phase with the changes in source voltage.

As indicated, voltage and current are not in phase with each other. Because of circuit inductance, they will always be slightly out of phase. The reactance does not dissipate power but stores energy in the magnetic (inductive) field (see Figure 4–23). This stored power is returned to the circuit whenever E and I reach opposite polarity.

FIGURE 4-22 In a pure inductive circuit current lags voltage by 90 electrical degrees.

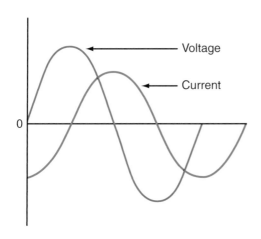

FIGURE 4-23 The result of current that is 90° out of phase is indicated by true power that has equal positive and negative pulses that average zero over one full cycle.

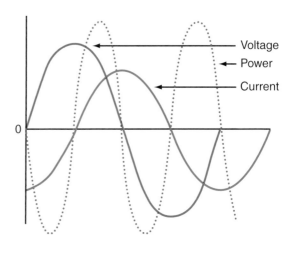

As discussed earlier, power stored by the reactive component of the circuit is referred to as Volt Amp Reactive (VAR)s. This is temporary stored power in the magnetic field of the coil. As the energy is stored, it is not consumed and is not measured as watts. The VARs are calculated by using the line current and the E calculation due to X_L, or using circuit current squared times the X_L. VARs are represented in the power triangle as the vertical components of the right triangle. The power triangle will be discussed after the components parts and power factor are identified.

4.7 Apparent Power

Apparent power can be computed by multiplying the circuit voltage ($V_{applied}$) by the total current flow (I_T) with no regard for phase angle. Remember that total circuit current is determined by combining the reactive and the resistive components. The total current would be the measured current, as it is the amount of current that is apparently being used by the circuit and therefore is a measurable quantity. For instance, in a circuit with 220 V and total current of 14 A, the volt-amps (VA) are:

$$VA = V_{applied} \times I_T$$

$$VA = 220 \times 14$$

$$VA = 3080$$

4.8 Power Factor

The power factor is not an angular measure but a numerical ratio. This is the ratio of true power to apparent power. If the ratio is 1, or unity, it means that the circuit is a resistive circuit. The opposite is also true: if the circuit has no resistive component and is entirely reactive, the power factor is 0. The power factor can be calculated by dividing the circuit resistive value by its similar total impedance value. Power factor can be calculated by dividing the voltage drop across all the resistors by the total circuit voltage, total resistance by impedance, or watts by volt-amps.

For example, given a true power of 1936 W with a VA of 3080, find the power factor (*PF*). To turn the decimal into a percentage, multiply the answer by 100:

$$\% \ PF = \left(\frac{W}{VA}\right) \times 100$$

$$\frac{1936}{3080} \times 100 = PF = 62.9\%$$

or

$$PF = VR, \text{ voltage drop across the resistor, divided by } V_T,$$
$$\text{the total voltage applied to the circuit.}$$

$$\% \ PF = \left(\frac{R}{Z}\right) \times 100$$

Power Triangles

To finish the circuit represented in Figure 4–16, we need to analyze the power triangle that represents the power values calculated. The true power is always represented on the horizontal line of the triangle. The VARs are represented on the vertical line of the right triangle, and the apparent power is the hypotenuse of the right triangle. The triangle is congruent to the impedance triangle and the voltage vector. The watts as previously calculated are 13.1 W, the VARs are 32.8 VARs and the apparent power is 35.4 VA, as depicted in Figure 4–20. The phase angle between the horizontal and the hypotenuse is calculated by the cosine of the angle, or adjacent side divided by the hypotenuse, using this formula:

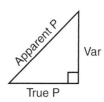

$$\%PF = \left(\frac{\text{True power}}{\text{Apparent power}} \right) \times 100$$

Power factor could also be calculated from the impedance triangle because the congruent angles are the same in both triangles so the formula would be:

$$\%PF = \left(\frac{\text{Resistance}}{\text{Impedance}} \right) \times 100$$

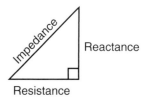

The voltage vector has the same angle so the formula would be:

$$\%PF = \left(\frac{\text{resistive volt drop}}{\text{applied line voltage}} \right) \times 100$$

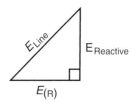

In all cases the power factor is 62.9%

VARIATIONS IN CIRCUIT VALUES

The series RL circuit will change based on variables. By using Table 4–1 you can see that changing the frequency, the henry value of the coil, or the resistance can change many circuit quantities and outcomes. Some of the quantities are not obvious as the shape of the triangles change and not all values change proportionally. Keep in mind that in a series resistor and inductor circuit, if only the resistance increases, true power will increase. And, if only resistance decreases, true power will decrease.

TABLE 4–1 Use this table to compare what will happen if you change one variable at a time in a series *RL* circuit. The top line identifies the calculated value, and the left column indicates which way the variable changes. Use the chart to predict the effects of different calculations as the variables are changed one at a time. Use the impedance, voltage, and power triangles to verify trends.

Variable	X_L	E_L	E_R	I_T	Z_T	P_A	P true	VARs	Phase angle	PF
Freq Inc	Inc	Inc	Dec	Dec	Inc	Dec	Dec	Inc	Inc	Dec
Freq Dec	Dec	Dec	Inc	Inc	Dec	Inc	Inc	Dec	Dec	Inc
Henry Inc	Inc	Inc	Dec	Dec	Inc	Dec	Dec	Inc	Inc	Dec
Henry Dec	Dec	Dec	Inc	Inc	Dec	Inc	Inc	Dec	Dec	Inc
Resis Inc	Same	Dec	Inc	Dec	Inc	Dec	—	Dec	Dec	Inc
Resis Dec	Same	Inc	Dec	Inc	Dec	Inc	—	Inc	Inc	Dec

Inc = increase; Dec = decrease

RISE TIME OF INDUCTOR CURRENT

When DC voltage is applied to a resistive load, as shown in Figure 4–24, the current will instantly rise to its maximum value. A resistor has a value of 6 Ω, and it is connected to a 12-V source. When the switch is closed, the current will rise instantly to a value of 2 A:

$$I = \frac{V}{R} = \frac{12}{6} = 2 \text{ A}$$

If an inductor is added to the circuit, as shown in Figure 4–25, and the switch is closed the current cannot change instantaneously because of the CEMF produced by the inductor. When the switch is first closed, the current tries to rise to its maximum value, just as it did in the circuit in Figure 4–24; however, the sudden change in current (from zero) will cause the inductor to induce a CEMF that opposes the current to immediately rise to the maximum value. Remember that:

$$V_{\text{CEMF}} = -L\left(\frac{\Delta I}{\Delta t}\right)$$

FIGURE 4–24 A DC circuit and associated graph of time versus current flow after switch closes.

FIGURE 4–25 An AC circuit with a 5 time constant charge graph for current flow after switch closes.

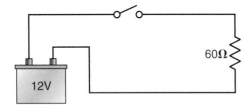

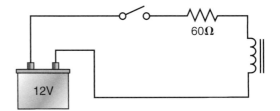

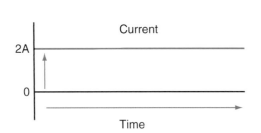

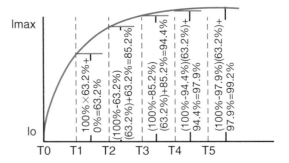

The CEMF opposes the rise in current and increases the amount of time required for it to reach maximum. As the current flow increases and approaches the maximum current flow value, the rate of change decreases, as does the amount of induced voltage. When the current reaches (or gets very close to) its maximum, the inductor no longer creates any CEMF; consequently, the maximum current in both circuits (Figures 4–24 and 4–25) is the same.

The amount of time required for the current flow to reach the maximum value is expressed in time constants. In a circuit with inductance and reactance, the time constant is defined from the following formula:

$$T = \frac{L}{R}$$

where:
T = the time constant in seconds
L = the circuit inductance in henrys
R = the circuit resistance in ohms

As the switch is initially closed, the circuit experiences the greatest amount of change in current flow. The circuit reaches 63.2% of the maximum current flow in the first time constant. As the circuit proceeds through the next time constant increases to 63.2% of the remaining current maximum, the next is another 63.2% of the remaining total and so forth. After five time constants, the current has reached 99.2% of the maximum and the increase is considered complete. The amount of change in current for each time constant remains the same percentage from 0% to 100% and the number of time constants (five) necessary to reach the maximum constant current will be consistent, regardless of the size of the inductor and the source voltage. The total time in seconds varies depending on the circuit values because the time constant in seconds varies. This type of current change is called an exponential change. The inductance affects the circuit during charge and discharge. Just as the circuit takes five time constants to reach a value very close to the maximum circuit current when the switch is closed, the same circuit will also take five time constants to reach a near-zero value once the switch is opened.

Take a look at Figure 4–26. Assume that switch S1 is closed and the current flow has reached its maximum value of 2 amps. If S1 is suddenly opened and S2 is simultaneously closed, the inductor magnetic field will start to collapse. As it does, the CEMF induced in the inductor will cause current to flow through the resistor and switch S2. Initially, the current will be 2 amps since the current through an inductor will not change instantaneously. Gradually, however, the current will decay, and it will do so in the same amount of time it took to rise to the final value.

FieldNote!

Rise times and decay time of the coil current are consistent. The result of this five-time constant consistency allows us to use this effect for timing circuits. Another reason to use the effects of the $\frac{L}{R}$ time is to protect a circuit from the voltage spike caused by a collapsing inductor magnetic field. If we allow the magnetic field to collapse quickly, there is a large voltage produced in the coil and that voltage may ruin other circuit components. If we place a resistor with low ohms in the circuit with the coil, the time constant becomes longer than with no resistor (very large resistance). The effect is to allow the current to decay over a longer period of time and reduce the voltage "inductive kick."

FIGURE 4–26   The decay time for current to return to zero is also 5 time constants.

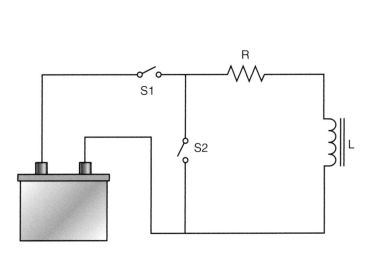

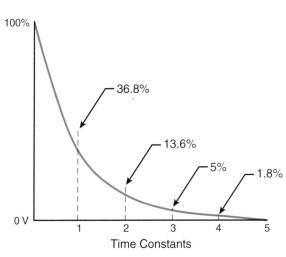

EXPONENTIAL RATES AND TIME CONSTANTS

The exponential curve describes a rate of occurrences in nature. The curve is divided into five time constants or divisions. Each time constant will have a change equal to 63.2% of the remaining difference or value. During the first time constant, the value will rise from 0% to 63.2% of its total value, during the second time constant it will rise to 63.2% of what is left to the maximum, and so on.

Example

It takes 5 seconds for the current to rise to a maximum 10 amps in a circuit. This means that the one time constant is one second. Table 4–2 shows the rise times and the total current for each time constant.

An important point needs to be made about this type of current change: Theoretically, the current will never reach the final value. You can see this by inspecting the values in Table 4–2. Notice that with each time constant, the total current gets closer to the final value. Notice also, however, that with each time constant, the change in current gets smaller. In this example, the current actually reaches only 99.3% of the maximum. Another time constant would make it closer but still not reach maximum.

TABLE 4–2   Five Time Constants

Time	Rise	Final Current
1	63.2% × 10 A = 6.32 A	6.32 A
2	63.2% × (10 A − 6.32 A) = 2.33 A	2.33 A + 6.32 A = 8.65 A
3	63.2% × (10 A − 8.65 A) = 0.853 A	0.853 A + 8.65 A = 9.5 A
4	63.2% × (10 A − 9.5 A) = 0.316 A	0.316 A + 9.5 A = 9.81 A
5	63.2% × (10 A − 9.81 A) = 0.12 A	0.12 A + 9.81 A = 9.93 A

SUMMARY

In this chapter, you practiced solving for the total inductance in a series circuit containing more than one inductor. You also learned how to find total inductive reactance in a series circuit containing more than one inductor. Each inductor has to be physically independent from other inductors in the circuit for these formulas to work; that is their magnetic fields must not interfere with each other. If the magnetic fields do interact, mutual induction formulas using the coefficient of coupling must be used. Like resistances in series, other circuit unknowns are calculated to make a complete circuit analysis. When inductors are connected in series, the total inductance is equal to the sum of all the inductors.

Once all the inductance and inductive reactance is determined the circuit must be calculated with all additional series resistors. The total opposition is called the impedance of the circuit. The impedance can be calculated using the Pythagorean theorem to determine the hypotenuse of the right triangle, or the same impedance can be calculated using trigonometry and finding the vector sum of the R and X_Ls.

The final impedance triangle will be congruent with the final voltage vector. The voltage across each component in the circuit is calculated by the total current multiplied by the total opposition of the component. All the voltages must add up (vectorially) to the total voltage applied. Along with the voltage triangle, the power triangle will also be congruent. The true power is measured in watts, the reactive power is measured in volt-amps-reactive (VARs), and the hypotenuse of the power triangle is the apparent power and is measured in volt-amps (VA).

Finally, the rise time and decay time of LR circuits were examined. It was determined that no matter what the henry value or the resistance value of the circuit is, the current always takes five time constants to reach maximum or decay to zero in the circuit. The time constant measured in seconds will depend on $\frac{L}{R}$.

REVIEW QUESTIONS

1. Three coils are connected in series. Assuming no mutual induction and the value of the coils are .3 H, .7 H, and 1H, respectively, what is the inductance of the total circuit?

2. Each of the coils in Question 1 is connected to a 120-V AC circuit at 60 Hz. What is the total inductive reactance of the circuit?

3. If two 5-H coils are connected in series and they have an aiding 50% coefficient of coupling, what is the total inductance of the circuit?

4. Explain the effect of an opposing coefficient of coupling on the circuit inductance.

5. Find the degree lag of the current in an AC circuit when a 100-Ω resistor is connected in series with a coil with 50 Ω of resistance and 150 Ω of reactance.

6. Find the circuit Z of a series circuit with two resistors, each 75 Ω, and a coil with 30 Ω of R and 100 Ω of reactance.

7. If a coil creates a 30° lagging current and the total Z is 50 Ω, what is the X_L?

8. Determine the voltage dropped across the coil and the resistor in the circuit of Figure 4–27.

FIGURE 4–27 A resistor and coil in series.

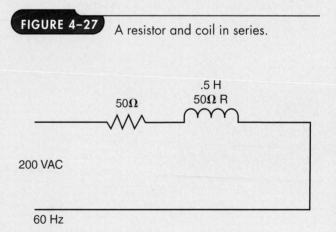

9. Find the total Z and the angle θ for the series circuit totals of Figure 4–28.

10. Compute the power factor for the circuit in Figure 4–28.

FIGURE 4–28   LR series circuit for Question 9.

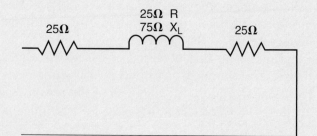

11. Explain what is meant by VARs.

12. How is apparent power calculated for an AC circuit?

13. How is power factor calculated when the phase angle is known?

14. Refer to Figure 4-28. Assume the voltage is 240 V AC. Calculate the true power, the apparent power, and the reactive power for the circuit totals.

15. How long does the current take to reach maximum after first energized, using a 2 H coil and a 1 KΩ resistor?

5

Capacitors, Capacitance, and Capacitive Reactance

O U T L I N E

OVERVIEW

So far, you have learned a great deal about resistance and resistive circuits, and you have learned the fundamentals of inductance and inductive reactance in series circuits. In this chapter, you will learn about the third type of passive element: the capacitor.

Inductance, as you learned, opposes a change in the current flow. It does this by creating a counter electromotive force when the current tries to change. Capacitance is the complement of inductance. Capacitors store an electrical charge. If the applied voltage is greater than the stored charge, current flows into the capacitor. If the capacitor's stored charge is greater than applied voltage, capacitive current flows back toward the source. If the capacitor charge is equal to the applied voltage, no current flows. The capacitor appears to oppose a change of voltage. Capacitors do have an opposition to current flow known as capacitive reactance.

Capacitors are used in electrical circuits of all kinds. They are used to block DC and pass AC, to correct power factor problems in systems, to set timing constants for various types of oscillators, and for a variety of other applications.

Generally, capacitor operation is not as intuitive as inductor operation. The formulas, concepts, and circuit analysis used when capacitors are involved are very similar to those involving inductors.

OBJECTIVES

After completing this chapter, you should be able to:
- Define capacitance
- Determine factors affecting capacitance based on physical characteristics
- Calculate capacitance based on formulas
- Identify types of capacitors and their uses
- Describe the causes of leading current in capacitive circuits
- Determine time constants and charge and discharge rates
- Determine capacitive reactance

CAPACITANCE

Capacitance refers to the capacity of a circuit or circuit component to store an electrical charge. This charge is in the form of voltage and the storage medium is the electrostatic field between the conducting plates of the capacitive device. As you have learned, inductors have a characteristic to oppose a change in current as it slows or impedes the current waveform. Capacitors have the characteristic to oppose a change in voltage. Inductors store energy in a magnetic field. Capacitors store energy in an electrostatic field. The effect of capacitors in an AC circuit counteracts the effects of inductors. The effects oppose each other.

Capacitance appears in many forms with various conducting surfaces and various insulators between the conducting surfaces. The closest analogies to a capacitive storage device would be static electricity stored as you move across a carpet or the buildup of charges in the cloud to ground lightning storm. There is an electrical charge of electrostatic energy built up with a normal insulating material between the two potentials. When the potential is too large or the insulation is too weak, the electrostatic field will discharge. We can use this capacity to hold a charge and not break down the insulating layer to our advantage, as we use the storage medium to control the voltage needs of the circuit.

Capacitance is measured in **farads.** They are used to block DC current and pass AC current. One farad of capacitance is measured when one volt applied to the capacitors' plates results in one coulomb of charge. Remember that one coulomb is a quantity of electrons. One coulomb is 6.28×10^{18} electrons. In other words, when one coulomb of electrons collects on one plate compared to the other plate, there is a one-volt difference between the plates. As you will see, the electrons do not pass through the insulating dielectric but do exert electrical pressure. One farad of a capacitance is a fairly large value, so submultiples of capacitance are often used. A common submultiple of the farad is the **microfarad,** represented as μF. The microfarad is .000001 farads, or 1×10^{-6} farads. Another common submultiple is the **picofarad,** also known as the **micro-microfarad** or 10^{-12} farads. This can be represented by pF or μμF. A less common multiple is the **nanofarad,** or 1×10^{-9} farads. This can be represented by nF.

5.1 Schematic Symbols and Markings

The symbol most used in schematics is a straight line opposed by an arc (Figure 5–1). There are other variations used at different times and in different circuit designs around the world.

FIGURE 5–1   Schematic symbols used for capacitors both fixed and variable.

Fixed Value Capacitor

Variable Value Capacitor

FieldNote!

The original capacitor was known as a Leyden jar. Around 1745 a group of professors at the University of Leyden in the Netherlands discovered that an electrical charge could be held on conducting surfaces if the surfaces were separated by an insulator (glass). Therefore the first capacitors were called Leyden jars. Michael Faraday did extensive research into the effects of electrostatic and electromagnetic field in the mid-1800s. The unit of measure for the capacity of a capacitor, the farad, was named in Faraday's honor.

Capacitor
A capacitor is an electric circuit element used to store charge temporarily. In general it consists of two metallic plates separated and insulated from each other by a dielectric. A capacitor may also be referred to as a condenser. (Excerpted from American Heritage Talking Dictionary. Copyright ©1997 The Learning Company, Inc. All Rights Reserved.)

Dielectric
A nonconductor of electricity, especially a substance with electrical conductivity less than a millionth (10^{-6}) of a siemens. A dielectric has the ability to insulate against a current flow between conducting surfaces. (Excerpted from American Heritage Talking Dictionary. Copyright ©1997 The Learning Company, Inc. All Rights Reserved.)

CAPACITANCE FORMULAS

The general formula for capacitance is: $C = \dfrac{Q}{V}$
where:
C = capacitor value in farads
Q = charge in coulombs
V = voltage applied

This formula is used to compare the charge a capacitor can hold per volt that is applied to the capacitor. To illustrate this concept, the larger the plates of the capacitor, the more electrons (measured in coulombs) the capacitor can hold for every volt of potential applied and, therefore, it has a larger capacity or capacitance.

Example

What is the capacitance of a capacitor if you have a charge of .1 coulombs when 100 volts are applied?

Solution: $C = \dfrac{Q}{V}$

$$C = \frac{.1}{100}$$

$$C = .001 \text{ F}\mu \text{ or } 1000 \text{ }\mu\text{F}$$

Example

What voltage would need to be applied to a 500 μF capacitor to get a 0.5 coulomb charge?

Solution: $C = \dfrac{Q}{V}$ transposed to $V = \dfrac{Q}{C}$

$$V = \frac{.05}{.0005 \text{ F}}$$

$$V = 100 \text{ V}$$

CAPACITOR CONSTRUCTION AND RATINGS

5.2 Physical Characteristics

A **capacitor** may perform a variety of jobs, such as storing an electrical charge to produce a large current pulse, providing voltage for timing circuits, or making a correction to power factor. Take a look at Figure 5–2. A capacitor consists of two conducting surfaces, usually referred to as plates, separated by some type of insulating material, called the **dielectric.** The plates are opposite electrical polarities (+, −) The insulating material can also use air as the dielectric. Capacitors are devices that appear to oppose a change in voltage.

Factors that affect the capacitance of the capacitor include the area of the plates, the distance between the plates, and the type of dielectric used and the number of capacitive actions. The formula for calculating the capacitance of a capacitor measured in farads is:

$$C = \frac{.225 \ AK}{10^6 \ d} \text{ in English units}$$

where:

C = capacitance in microfarads

A = cross-sectional area of one plate in square inches

K = dielectric constant of the material between the plates (for air, $K = 1$)

d = the thickness of the dielectric (or the distance between the plates) measured in inches

or

$$C = \frac{(8.85 \ AK)}{d \times 10^{12}}$$

where:

C = farads

A = area of the plates in square meters

K = dielectric constant

Because capacity of a capacitor is based on physical characteristics, like the inductance of an inductor, we can compute the capacity from the formula above, without any circuit conditions needed.

Capacitor basic construction with plates and dielectric.

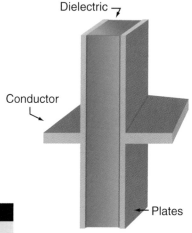

Dielectric

Conductor

Plates

Example

Find the capacitance in farads for a capacitor with the following information: A two-plate capacitor has a dielectric of glass with a K of 10. The plate area that interacts with the other plate is .02 square meters. The distance between the plates is .001 meters.

Solution:

$$C = \frac{(8.85 \ AK)}{d \times 10^{12}}$$

$$C = \frac{(8.85 \times .02 \times 10)}{.001 \times 10^{12}}$$

$$C = \frac{1.77}{10^{9}}$$

$$C = .00000000177\text{F} \ or \ 1.77 \ \text{nF}$$

Example

Find the microfarad value of a capacitor with the following dimensions: There are two plates separated by an insulating oil dielectric with a K of 4. The plates are 3 square inches each and the distance between the plates is .2 inches.

Solution:

$$C = \frac{.225 \ AK}{10^{6} \ d}$$

$$C = \frac{.225 \times 3 \times 4}{10^{6} \times .2}$$

$$C = \frac{2.7}{2 \times 10^{5}}$$

$$C = \frac{2.7}{200,000}$$

$$C = .0000135 \ \text{F} \ or \ 13.5 \ \mu\text{F}$$

TechTip!

The capacitance formulas appear in many different forms in many different publications. The two versions are for English units and for metric units. You may see the exponents in various mathematic variations such as multiplying by 10^{-12} or dividing by 10^{12}. Some formulas keep the scientific notation intact while others use the exponents to convert directly to farads or microfarads. Be sure you know which factors are in which units of measure for the correct formula.

THEORY OF CAPACITIVE ACTION

To charge a capacitor, you connect it to a DC power source and electrons will be removed from the side connected to the positive polarity and deposited on the side of the capacitor connected to the negative polarity (Figure 5–3). This accumulation of electrons will continue until the voltage across the capacitor equals the voltage across the source. When these two voltages become equal, the current will stop, and the capacitor is charged. A good rule to know about capacitors and current is that current movement can occur or change only during the time when the capacitor is charging or discharging. Since there is no "flow" of electrons through the capacitor because the plates are separated by a dielectric (insulator), the current just seems to move through the conductor. As an example see Figure 5–4. The current flows into the terminal of the capacitor and begins to fill the top plate with electrons, giving that plate a negative charge. The current appears to flow out of the bottom plate that is connected to the positive electrical source and it draws electrons away from the bottom plate, giving it a positive charge. From the circuit perspective the current appears to flow through the capacitor, but it does not. The electrons collect on the top plate and leave the bottom plate, which creates a potential difference between the plates. As this charge builds, the potential developed by the capacitor builds to eventually be equal and opposite to the applied source voltage. At this point the capacitor is fully charged and the circuit current ceases.

FIGURE 5–3   The charging circuit of a capacitor with a capacitor charging voltage curve.

FIGURE 5–4   Current flows to the capacitor but not through the insulator.

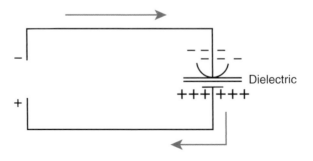

FIGURE 5-5   An analogy of a bathtub and a capacitor with a water system.

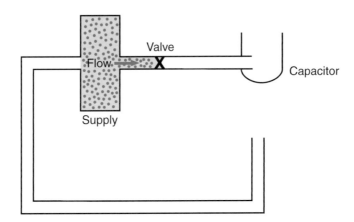

In theory, the capacitor should remain charged forever if the source voltage is removed. In actual practice, it will not. No dielectric is perfect. The electrons will eventually move from the negative plate to the positive plate or "leak through" the dielectric, causing the capacitor to discharge and become electrically neutral. This current flow is called leakage current and is inversely proportional to the resistance of the dielectric and directly proportional to the voltage across the plate. If the dielectric of a capacitor becomes weak, it will permit an unacceptable amount of current flow. Such a capacitor is called a leaky capacitor.

ELECTROSTATIC CHARGE

Capacitance is the ability of a circuit component to store an electrical charge. In previous chapters, you learned that a coil has a constantly moving electromagnetic field. A capacitor, on the other hand, has a changing electrostatic field. The term *electrostatic* refers to the fact that it has an electrical charge that is stationary. This is the charge that is produced when the electrons are removed from one plate and deposited on the other. This storage of a potential difference, or voltage, causes tension and a pull on the electrons in the insulating material. Think of this as a slingshot being drawn back. When the two sides of the capacitor are shorted or have a circuit, the electrons are "fired," and the tension between the two sides of the capacitor is released, returning the dielectric material to neutral (Figure 5–6).

FIGURE 5-6   The electrostatic field of a charged cap versus an uncharged capacitor.

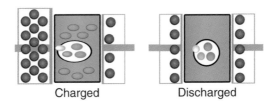

Charged   Discharged

VOLTAGE AND CURRENT RELATIONSHIP IN A CAPACITOR

You previously learned that the relationship between the voltage drop on an inductor is given by the formula:

$$E_{\mathrm{L}} = -L \times \left(\frac{\Delta I}{\Delta t} \right)$$

where:
E_{L} = the voltage across the inductor
L = the inductance in henries

You also learned that the term $\dfrac{\Delta I}{\Delta t}$ is the change in current with respect to the change in time. A capacitor has a very similar formula that relates its voltage and current:

$$I_{\mathrm{C}} = C \left(\frac{\Delta E_{\mathrm{C}}}{\Delta t} \right)$$

where:
E_{C} = the voltage across the capacitor
I_{C} = the current through the capacitor
C = equals the capacitance in farads
$\dfrac{\Delta E_{\mathrm{C}}}{\Delta t}$ = the change in capacitor voltage with respect to the change in time.

Both of these formulas will be very important as you learn about the phase shifts associated with capacitors and inductors.

We can now elaborate on the statement that is used throughout the electrical occupations: Inductors oppose a change in current and capacitors appear to oppose a change in voltage. Inductors create a CEMF to oppose current flow and capacitors build a voltage to oppose the circuit-applied voltage.

CAPACITORS AND CURRENT

Remember that a capacitor is made from two metal plates separated by an insulating material called a dielectric. The dielectric prevents the flow of electrons through the capacitor. When connected to alternating voltage supply, current appears to flow through the capacitor. The reason this appears so is that in an AC circuit, the polarity is continually changing, and the current is changing direction. This causes the capacitor to continuously charge and discharge.

To demonstrate, look at Figure 5–7. The capacitor is represented by a hollow tube of metal. At one end of the tube is a stretchable membrane that flexes but does not allow material to flow through. As the golf balls are pushed into one end a golf ball exits the other end. As more golf balls are added to the left side more golf balls exit the right side. Since the golf balls all look the same, it appears that the balls are moving through the tube, but they are not. In fact, the membrane is stretching more and more until no more balls can enter the tube. The capacitor is full but no balls have actually moved through the capacitor.

A representative capacitor with apparent current flow through the tube but no real current through the tube.

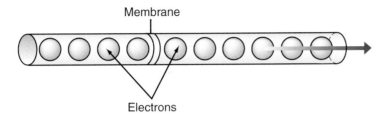

Membrane

Electrons

Before we analyze the capacitor voltage and current relationship in an AC circuit, we need to first look at it from a DC vantage point. As the capacitor is charged—or filled—from the DC source, the voltage increases on the capacitor plates and eventually will oppose any further current flow. As we look at this process, we can see the voltage build across the capacitor. Initially when the capacitor is empty of charge it has zero volts across it, as represented in Figure 5–8. As the current begins to fill the capacitor, the cap voltage increases and likewise the current begins to decrease. As we watch the capacitor, the voltage increases as the current diminishes to zero as at point 1. When we compare the graphs of this occurrence as in Figure 5–8 it appears as though the current "through" the capacitor is 90° ahead of the voltage of the capacitor. Now, if we will connect the voltage source to the capacitor in the opposite direction, we will watch the current flow out of the cap in the opposite direction, filling the bottom plate with electrons and while watching the voltage across the cap equalize to zero as at point 2. If we allow the capacitor to continue to charge, we again see the current going from peak toward zero as it flows to the capacitor in the opposite direction, and the voltage increases from zero to maximum at point 3 in Figure 5–8. This again makes it appear as if the current leads the voltage through the cycle. To complete the process, we will reconnect the DC source back to the original polarities. The current goes from zero to maximum in the opposite direction until the voltage across the capacitor is again zero at point 4. This completes an alternation of applied voltage. The current in the capacitor circuit leads the voltage across the capacitor by 90 electrical degrees.

FieldNote!

"ELI the ICE man" is a phrase used in the industry to remind practitioners of the current and voltage relationships that exist in inductive circuits represented by L and in capacitive circuits represented by C. E leads I in an inductive circuit, hence the acronym ELI. This is the same circumstance as I lagging E as we have previously learned. The acronym ICE indicates that I leads E in a capacitive circuit. Thus, "ELI the ICE man" is a mnemonic used to help memorize the relationships of E and I in various circuits. Remember that in a purely resistive circuit, the E and I are in phase and have synchronous waveforms. In AC circuits with L and C and R the current with reference to the voltage may be anywhere from 90° leading to 90° lagging, or in between.

An illustration of how current leads voltage in a capacitor.

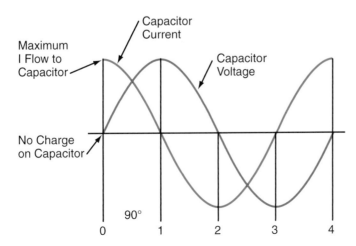

Capacitor Current

Maximum
I Flow to
Capacitor

Capacitor
Voltage

No Charge
on Capacitor

90°

0   1   2   3   4

VOLTAGE RATINGS

Voltage rating of a capacitor is the amount of voltage differential that the dielectric can withstand before puncturing or breaking down. The rating has to do with the dielectric strength or volts per mil multiplied by the mil-inches thickness of the dielectric. As the electric field intensity increases as voltage is applied, the orbits of the electrons stored on the plates actually distort and finally break through the dielectric. The electron flow punctures the insulator and creates an arc through, making the dielectric no longer useful. The overall voltage rating of the capacitor is based on the dielectric strength of the insulating material, measured in volts per mil-inch or volts per .001 inches multiplied by the thickness of the insulator in mils. Some example dielectric strengths are listed in Table 5–1.

TABLE 5-1	The Approximate Values of Dielectric Strength in Volts/Mil Inch
Kind of Material for Dielectric	Approximate Dielectric Strength in Volts per Mil-inch (.001")
Air	≈80
Glass	≈200–2000
Wax paper	≈1200
Mica	≈2000
Ceramic	≈500–1000

Example

What is the voltage rating of a capacitor with an air dielectric that is 5 mils thick?

Solution:

$$\text{Volts per mil} \times \text{mils thickness} = \text{voltage rating}$$

$$\left(\frac{\text{volts}}{\text{mil}}\right) \times \text{mils} = \text{voltage}$$

$$\frac{80 \text{ V}}{\text{mil}} \times 5 \text{ mils} = 400 \text{ V}$$

The voltage rating is extremely important for the life of the capacitor and should never be exceeded. The voltage rating indicates the maximum amount of voltage that the dielectric is intended to withstand without breaking down. If the voltage becomes too great, the dielectric will break down, allowing current to flow through the capacitor between the plates. In this condition, the capacitor is referred to as being shorted. Since the AC voltage rating is a root mean square (RMS) value, the actual peak value of the applied voltage on the capacitor will be considerably higher. You can calculate this value by multiplying the AC voltage by an RMS conversion value of 1.414.

For example, a 28-V AC value applied to a capacitor would require the capacitor to have at least a peak voltage rating of 28 × 1.414 = 39.59 V.

Capacitors designed for use in AC circuits are considered non-polarized capacitors, meaning they have no specific voltage polarity assigned to either plate. They operate the same way as in the previous discussion in that they store a charge of electrons on the plates for one half cycle at a time. The voltage applied to the capacitor is still determined by how much voltage can be applied in either direction before the dielectric breaks down. If the voltage rating on the capacitor is designated in AC volts, it is expressed as the RMS voltage that can be safely applied.

Some capacitors have their voltage rating expressed as Working Volts Direct Current, abbreviated WVDC. This tells you that the maximum voltage that the capacitor can withstand is listed in DC terms. If you want to know the maximum AC voltage that can be applied to this non-polarized capacitor, you must multiply the AC RMS value of the connected circuit by 1.414 to calculate the peak of the voltage. The peak AC voltage cannot exceed the WVDC without destroying the capacitor. AC electrolytic capacitors are not polarity sensitive as they are non-polarized.

Polarized capacitors have leads marked to indicate the positive or negative lead. These are known as DC electrolytic capacitors and cannot be used in AC circuits. In this case the voltage is marked in DC or DCWV.

5.3 The Effects of Dielectric Thickness

Dielectric thickness, or the spacing between the plates, affects the capacitor in two ways. As the spacing between the plates increases, the capacitance decreases as is verified by the formula: $C = \dfrac{.225\ AK}{10^6\ d}$. As "d" increases, the C value decreases. Imagine a capacitor plate separated by 10 feet of "air dielectric." There is virtually no capacity to that situation. As the conducting plates get closer the electrostatic field between them increases and the capacity increases. We have found ways to create a very small distance between the plates yet maintain a high degree of electrical insulation to the dielectric by using different insulating materials. As discussed previously, the voltage rating of the capacitor also depends on the type of dielectric and the thickness of the insulating material, as you can tell by the formula for voltage rating. You will note that a larger dielectric thickness decreases the C while increasing the voltage rating (Figure 5–9). The manufacturer of the capacitor determines the dielectric material. The material determines the dielectric constant and the subsequent thickness of the dielectric determines the dielectric strength.

FIGURE 5–9   As the distance between the plates changes it changes both capacitance and voltage rating.

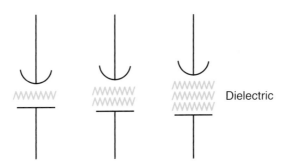

FIGURE 5–10

The effects of dielectric stress in a capacitor creates an elliptical orbit.

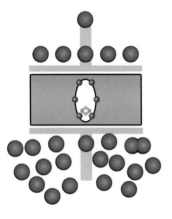

DIELECTRIC CHARACTERISTICS

5.4 Dielectric Stress

When a capacitor is charged, a potential exists between its plates. The plate with the lack of electrons has a positive charge, and the plate with the excess has a negative charge. The molecules in the dielectric are like small electrical dipoles; that is, they are more positive on one side than they are on the other. The negatively charged plate attracts the positive side of the molecules, and the positively charged plate attracts the negative side of the molecules. This attraction causes the molecules in the dielectric to twist and stretch. They do not actually move to any great extent, but, like a strong spring or rubber band, they will store energy. This action stores energy in the dielectric of a charged capacitor. It has little effect on the capacitor during AC operations; however, it does have some ramifications during DC testing of insulation (Figure 5–10).

5.5 Dielectric Constants

The type of dielectric used is an important factor in determining the amount of capacitance a capacitor will have. Materials are assigned a number called the dielectric constant, usually abbreviated with an uppercase "K" or the Greek letter epsilon (ϵ). Air is assigned the number 1 and is used as the reference point for comparison. Table 5–2 shows the dielectric constants for various materials. A change in the dielectric of the capacitor will change its rating. For instance, a capacitor with an air dielectric (1) and a capacitor rating of 5 μF would change its rating if the dielectric changed to Teflon (2):

$$5 \text{ μF} \times 2 \text{ (dielectric rating)} = 10 \text{ μF}$$

TABLE 5–2   Dielectric Constants for Typical Dielectric Materials

Material	Dielectric Constant
Air	1
Bakelite	4–10
Castor oil	4.3–4.7
Cellulose acetate	7
Ceramic (titanium dioxide)	10–110
Ceramic (barium-strontium titanate)	up to 7500
Dry paper	3.5
Hard rubber	2.8
Insulating oils	2.2–4.6
Lucite	2.4–3.0
Mica	6.4–7.0
Mycaflex	8.0
Paraffin	1.9–2.2
Porcelain	5.5
Pure water	81
Pyrex glass	4.1–4.9
Rubber components	3.0–7.0
Teflon	2

Figure 5–11 shows some different types of commercial capacitors that you may encounter during your electronics career.

FIGURE 5–11   Photo of oil filled and electrolytic capacitors.

NUMBER OF PLATE INTERACTIONS

Many capacitors are made up of long plates that are rolled into a cylinder as shown in Figure 5–12. In the formula $C = \dfrac{.225\ AK}{10^6\ d}$, we assume there are only two plates. If there are more than two plates, as in Figures 5–13 and 5–14, the basic formula can be altered. For instance, look at Figure 5–13 and you will see that there are three positive plates and three negative plates. The area of interaction is only where the plates overlap, so the actual area of the plates is only 2 square inches. However, the number of interactions increases to 6 locations where a negative plate reacts to a positive plate. Count the actual number of plates as 7. The number of interactions is one less than the number of plates or number of plates -1 ($n - 1$) for flat plates. Rolled plates have an advantage in that the top positive plate reacts with the negative plate

FIGURE 5–12   Rolled plates of capacitors have large plates in small physical space.

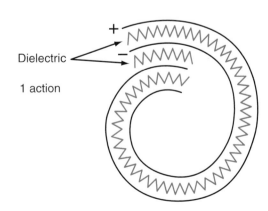

Dielectric

1 action

FIGURE 5-13 Capacitor with multiple overlapping plates increases plate interactions.

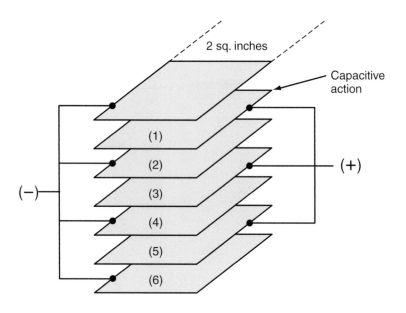

FIGURE 5-14 Rolled capacitors with fixed plate area but multiple dielectrics increase plate interaction.

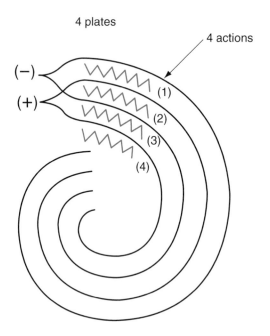

twice as it is rolled. The number of actions for an even number of rolled plates is the same as the number of plates. If the number of plates is odd, the number is again $n - 1$. There is no advantage to rolling an odd number of

plates. Now we can modify the basic formula to: $C = \dfrac{.225 \, AK \, (n - 1)}{10^6 \, d}$ for flat plates and odd number of rolled plates. This formula is universal when using English units as we measure the actual overlap of plates as the area of plate and the $(n - 1)$ is the number of plates -1 to indicate how many times they interact. A 2-flat-plate capacitor will still have only an interaction of one. From this example we see that we can reduce the size of each plate but increase the number of plate interactions and still produce the same capacitance as in Figure 5–15.

FIGURE 5–15 Two large flat plates have the same capacitance as four plates that are half the size.

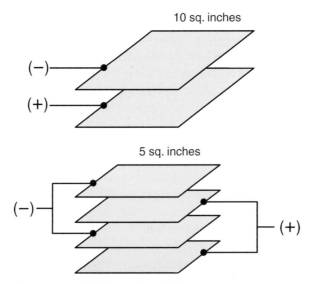

10 sq. inches

(−)

(+)

5 sq. inches

(−)

(+)

Example

Find the capacity and the voltage rating of a capacitor with the following construction: There are 9 plates and each has 3 square inches of interactive surface. The dielectric is mica. The plates are separated by .001 inches.

Solution:

$$C = \frac{.225\ AK\ (n - 1)}{10^6\ d}$$

$$C = \frac{(.225 \times 3\ \text{Sq''} \times 7 \times 8\ \text{actions})}{10^6 \times .1}$$

$$C = \frac{37.8}{10^5}$$

$$C = .000378\ \mu\text{F or } 378\ \text{pF}$$

The voltage rating is:

$$\text{Voltage rating} = \text{volts per mil} \times \text{mils thickness}$$

$$V = \frac{2000}{1\ \text{mil}}$$

$$V = 2000\ \text{V}$$

FIGURE 5-16

A variable capacitor was used to tune circuits with variable area of the plate interaction.

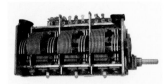

CAPACITOR TYPES

There are general classifications of capacitors. The variable capacitor is sometimes thought of as the adjustable cap that was used to tune radio frequency circuits such as in Figure 5–16. The plates were set so the area that interacted with the opposite polarity plates could be adjusted to increase or decrease the effective plate area. As the plate area increased, the C increased; similarly when there was less plate interaction the C decreased. In this example the dielectric is air. The formula $C = \dfrac{.225 \ AKN - 1}{10^6 \ d}$ confirms that by keeping the distance between the plates the same but changing the effective area of plates (Formula A) the capacity will change.

Another method of creating a variable capacitor is to adjust the distance between the plates. This method of adjusting the capacitance uses a small screw to physically move one plate closer to, or farther away from, a stationary plate. The capacitor style is referred to as a "trimmer" capacitor. As the screw is screwed in, the distance between the plates is reduced, decreasing "d" in the formula and increasing the capacitance. Conversely, as the distance between the plates is increased by turning the screw out, the capacitance decreases.

The second classification is the fixed capacitor. In this case the capacitance is determined by the manufacturer using all the parameters previously discussed. Once the manufacturing is complete, the capacity is determined and the capacitor is marked.

Other classifications of caps refer to the method of connection to the circuit. Polarized capacitors are capacitors that have definite positive and negative connection points for use in DC circuits. Nonpolarized capacitors can be used in DC as well as AC circuits without regard to the electrical polarity connected, which is imperative for AC circuits.

5.6 Electrolytic

FIGURE 5-17

Examples of polarized and non-polarized capacitors.

Electrolytic capacitors are used in many applications. They have a large capacity compared to their size. They are typically polarity sensitive meaning they have a positive and a negative terminal, but are also found in AC electrolytic values. As shown in Figure 5–17, the aluminum can housing forms one electrode and the other plate is aluminum foil rolled into the can. The dielectric is created when a current is passed through the capacitor when the foil is immersed in an electrolyte of a borax solution. As the current flows, it creates a very thin layer of aluminum oxide on the plates. This aluminum oxide is the dielectric and is extremely thin. Because it is so thin, the capacitance values can be very high compared to the space required. However, as you have learned, the voltage rating is decreased proportionally to the thickness of the dielectric. AC electrolytics again have high capacity for space required but the dielectric is formed from two directions on the plates so that AC can be applied without damaging the dielectric. These are routinely used as starting capacitors on motors.

FIGURE 5–18 Examples of tantalum, mica caps, and ceramic capacitors.

TechTip!

Ceramic, tantalum, and film capacitors are most commonly used in electronic circuits (Figure 5–18). The theory of how they are used is exactly the same as the theory you have learned; however, the small size and specific applications will be best discussed in a text on electronic applications.

5.7 Oil-Filled Capacitors

As the name implies, oil-filled capacitors are filled with an insulating oil that works as the dielectric. The plates are typically rolled aluminum and are separated by a paper saturated in oil. They typically have lower capacity than electrolytics for the same physical size, but have higher voltage ratings. This type of capacitor is used in motor duty, typically as a running capacitor.

TESTING CAPACITORS

Capacitors can be tested in a variety of ways. The basic theory of testing and knowledge of how a capacitor operates should guide you in the application of the test. Capacitators need to be tested for opens, shorts, or leaky conditions. As an example, we will discuss how to test an AC electrolytic capacitor that may be the cause of motor starting problems. To do this we will use a meter set to ohms. *Be sure power is off, the cap is disconnected, and that the cap does not have a stored charge.* Use an analog meter (easiest) and connect the ohmmeter to the capacitor. The meter should immediately go to very low ohms, indicating a large current flow from the battery in the meter to the capacitor. As you leave the meter connected the meter will indicate more and more ohms as the capacitor charges and current flow decreases. Finally, when the capacitor is fully charged the current flow ceases and the meter indicates infinite ohms. If this happens, the capacitor is okay. A shorted capacitor will have a large current flow and will constantly indicate very low ohms. A leaky capacitor will charge slightly—indicated by a moderate ohm reading—but will never go to infinite. It is leaking some current through the dielectric.

Once these basic tests give a general indication of the capacitor's status you can also determine its capacity value by measuring the AC current and the voltage across the capacitor. The capacitive reactance is determined and the capacitance can be calculated from that reactance value. This will be discussed later in this chapter. Another common method to test capacitance is done directly with a digital meter. Many digital meters such as that shown in Figure 5–19 have capacitance measurements built into the meter functions.

FIGURE 5–19 Many newer digital multimeters have a built-in capacitance check feature.

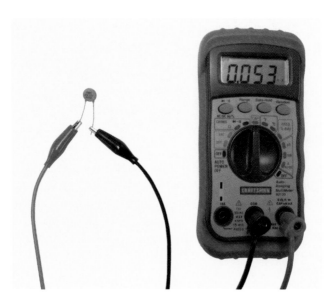

FIGURE 5–20

Capacitor being charged with a 10 VDC supply.

CHARGE AND DISCHARGE RATES

Capacitors (connected with resistors OR connected in series with resistors) charge and discharge at an exponential rate, the curve of which is divided into five time constants. During each time constant, the voltage will change by 63.2% of the amount left to reach the fully charged state. During the first time constant, the voltage across the capacitor will reach 63.2% of the supply voltage connected across it. During the next time constant, the capacitor will reach a higher voltage equal to 63.2% of the voltage left between the first time constant voltage and the supply voltage, which is then added to the voltage reached in the first time constant.

For example, assume that the capacitor in Figure 5–20 requires 10 seconds to reach the supply voltage of 20 volts. The capacitor can charge only to the system supply voltage. Table 5–3 shows the voltages across the capacitor during the 10 seconds.

As you can see, by the end of the fifth time constant, the voltage across the capacitor has reached 99.3% of the total voltage. If this continues past the fifth time constant, the voltage will continue to get closer to the final value; however, it will never reach the maximum. Notice that the voltage charge on the capacitor behaves in exactly the same way as the current flow through an inductor. Figure 5–21 shows the entire curve.

TABLE 5-3	Table Showing the Calculation for an Exponential Charge of a Capacitor	

Time Constant	Calculation	Final Voltage
1	63.2% of 20 V = 12.64	12.64 V
2	(20 − 12.64) × 63.2% = 4.65	12.64 + 4.65 = 17.29 V
3	(20 − 17.29) × 63.2% = 1.71	17.29 + 1.71 = 19.00 V
4	(20 − 19.00) × 63.2% = .632	19.00 + .632 = 19.632 V
5	(20 − 19.632) × 63.2% = .232	19.632 + .232 = 19.864 V

FIGURE 5-21	Voltage charge time curve based on the time constants.

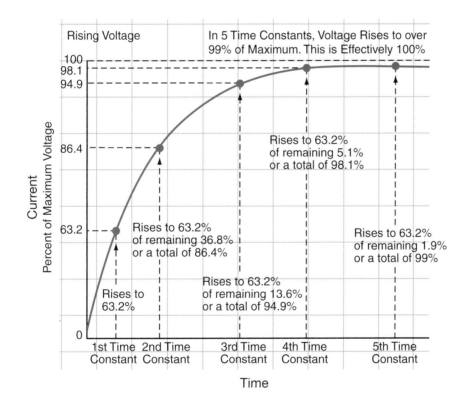

The capacitor discharges in the same manner. At the end of the first time constant, the voltage will decrease by 63.2% of its charged value. The voltage will continue to drop until it reaches approximately zero, when five time constants have elapsed (Figure 5–22).

Knowing that five time constants are required to reach the maximum charge is only half the picture. Of equal importance is determining how long a time constant is in seconds. The length of one time constant is dependent

FIGURE 5-22   Voltage discharge curve based on the number of time constants.

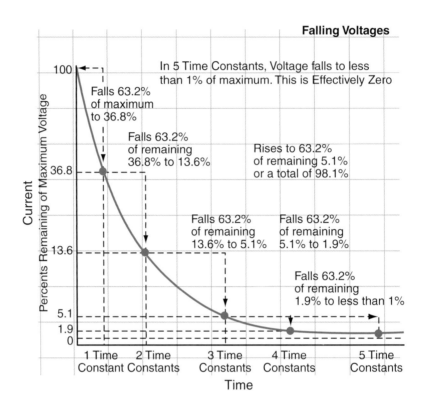

on the size of the capacitor and the resistance in series with the capacitor, and is given by the formula:

$$\tau = R \times C$$

where:
τ = one time constant in seconds
R = circuit resistance in ohms
C = circuit capacitance in farads

The lowercase Greek letter tau (τ) is used to designate the time required for one time constant. The capital letter "T" can also be used to indicate time.

Example

In a circuit with a 10 kΩ resistor and a 20 μF capacitor, how long is one time constant? How long will it take for the capacitor to fully charge?

Solution:

$$T = R \times C$$
$$T = 10 \text{ k}\Omega \times 20 \text{ μF} = 0.2 \text{ seconds}$$

To completely charge
$$T \text{ total} = 5 \times T = 1 \text{ second}$$

Example

If a 50 microfarad capacitor is fully charged to 100 volts and a 10 kΩ resistor is connected to the discharge path, how long will it take to discharge to approximately 5 volts?

Solution:

$$T = R \times C$$

$$T = 10 \text{ k}\Omega \times 50 \text{ }\mu\text{F} = 1.5 \text{ seconds}$$

Referring to Figure 5–22, 5 volts remaining on the cap is about 5% of the initial voltage. This corresponds to three time constants of discharge. $4\,T = 4 \times .5$ seconds or 2.0 seconds to discharge to nearly 5 volts

CAPACITIVE REACTANCE

To better explain the effects that a capacitor has on an AC circuit, we can look at a circuit with one-half of the AC sinewave applied. In Figure 5–23 we apply DC to illustrate the effect of charging the capacitor during a one-half cycle. This illustration shows that as the capacitor becomes fully

FIGURE 5–23 As the capacitor charges the voltage, the capacitor opposes the line voltage and reacts to reduce line current, a reaction to the capacitance.

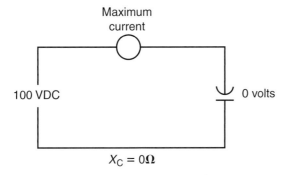

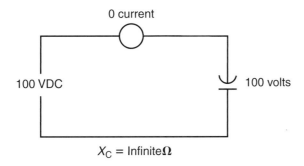

charged, the current in the circuit goes to zero. The opposition increases from zero opposition at first, to maximum opposition when fully charged. This effective opposition of the charged capacitor to the current in the circuit is identified as an ohmic value. The opposition is caused by the reaction of the capacitor in the circuit. The opposition to the current is called **capacitive reactance**. Although there is no capacitive reactance in a DC circuit, it helps visualize what happens as the capacitor continually charges and discharges one-half cycle at a time.

You have learned that this is similar to the opposing voltage, or countervoltage, in an inductive circuit caused by an inductor. In a perfect capacitor the current leads the voltage by 90 electrical degrees. Capacitive reactance, like inductive reactance, is measured in ohms but the symbol for reactance is X_C and again is measured in ohms. The X_C is added to other components by vector addition. Because the current actually leads the voltage, the X_C is vectored exactly opposite to the inductive reactance, or 90° down from the horizontal (Figure 5–24). The vector diagrams in Figure 5–24 are used to compare with the vector diagrams of a circuit with X_L. Although there is some internal resistance to the capacitor, it is typically disregarded because it is so small compared to the reactance.

Capacitive reactance
The opposition to AC current flow caused by a capacitor. The symbol for capacitive reactance is X_C. The formula to calculate capacitive reactance is $X_C = \dfrac{1}{2 \pi f C}$

where
X_C = ohms of reactance
f = frequency of the applied voltage measured in hertz
C = capacitance in farads

FIGURE 5–24   The vector for ohms of capacitive reactance is vectored down from the horizontal reference.

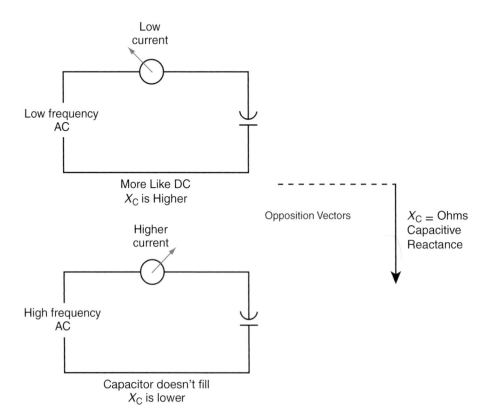

5.8 Capacitive Reactance Formula

Frequency Effects on Capacitive Reactance

Frequency effects on capacitive reactance create a condition where the capacitive reactance in a circuit decreases as the frequency of the circuit increases. You will examine this based on the reaction of current flow in a capacitive circuit and how that current flow would be an indicator of the capacitive reactance. As you recall, current is the rate of flow of electrons. If you connect a capacitor and a resistor to an AC circuit and measure the current that flows with a 30 Hz source compared to a 60 Hz source for the same capacitor and resistor and the same voltage level, you will find twice the current at 60 Hz compared to 30 Hz. This means the X_C is inversely proportional to the frequency. Referring to Figure 5–25, you can see that, for a given time period, the capacitor would charge to maximum, discharge to zero, charge to maximum in the reverse direction, and then discharge back to zero in one cycle of the 30 Hz waveform. This would produce a specific amount of AC current flow. Now consider the 60 Hz waveform. You can verify that for the same time period, the capacitor would go through two complete cycles of the same charge and discharge cycle. Therefore, for the same time period, twice as much current appears to flow in the circuit. If the current is twice as much with the same components and the same voltage level, the opposition to the current at 60 Hz must have gone to one half the original opposition at 30 Hz. The formula for verifying this effect is:

$$I = \frac{Q}{t}$$

where:
  I = current flow
  Q = charge in coulombs
  t = time in seconds

 FIGURE 5–25 Effects of increased frequency on the charge cycle of a capacitor.

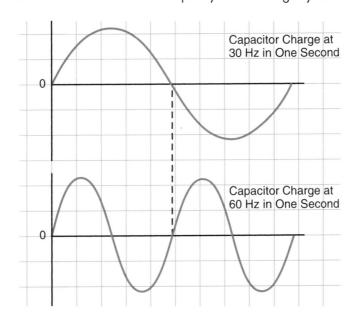

TechTip!

Although this is the recognized formula for X_C, it assumes the unit of measure for capacitance is in farads. As discussed previously, the farad is a large unit of measure and we would rarely see a farad capacitor. The microfarad is a more common unit, especially for electricians, and we can more conveniently use an altered formula for ease in calculations. The formula $X_C = \dfrac{10^6}{2\pi\, fC_{farads}}$ is much easier to use when the capacitor is already in microfarads (μF). A simplified version would be $X_C = \dfrac{159{,}000}{fC_{farads}}$.

Using Ohm's law you know that with the voltage the same in both situations, the current and the opposition are inversely proportional.

A more common formula for calculating the X_C in ohms is based on the size of the capacitor in farads, the frequency of charge and discharge, and the constant for a sine wave value is expressed as:

$$X_C = \frac{1}{2\,\pi\, fC_{farads}} \quad \text{or, more simply,} \quad X_C = \frac{159{,}000}{fC_{farads}}$$

where:
X_C = capacitive reactance
π = 3.14159
f = frequency in hertz
C = capacitance in farads

Example

A capacitor circuit has a 35 μF capacitor connected to a 48-V, 60-Hz line. How much current will flow through this line?

Solution:
First find the capacitive reactance. The C in the formula is in farads, so the conversion from μF to farads is multiplied by 10^{-6}. Thus, the line has 35×10^{-6} farads:

$$X_C = \frac{1}{2\pi\, fC}$$

$$X_C = \frac{1}{[2 \times 3.14159 \times 60 \times (35 \times 10^{-6})]}$$

$$X_C = \frac{1}{0.013195}$$

$$X_C = 75.79\ \Omega$$

If we use the equivalent formula mentioned in the previous sidebar, the equation solution is:

$$X_C = \frac{10^6}{2\pi\, fC_{farads}}$$

$$X_C = \frac{10^6}{2 \times \pi \times 60 \times 35\ F}$$

377 can be substituted for $(2\,\pi f)$ when calculating at 60 Hz:

$$X_C = \frac{10^6}{377 \times 35}$$

$$X_C = \frac{10^6}{13{,}195}$$

$$X_C = 75.786\ \Omega\ or\ 75.79\ \Omega\ \text{same as original formula}$$

Now use Ohm's law to calculate the current:

$$I = \frac{E}{X_C}$$

$$I = \frac{48\ V}{75.79\ \Omega}$$

$$I = 0.63\ A$$

CALCULATING CAPACITANCE FROM CAPACITIVE REACTANCE

Now that you have calculated capacitive reactance, you can determine the capacitance of any capacitor. Note that capacitance is also dependent on frequency and X_C. The formula is:

$$C = \frac{1}{2\pi f X_C}$$

Example

A capacitor is connected to a 220 V, 60 Hz line. The ammeter on the line indicates 3.2 A. What is the capacitance value of the capacitor?

Solution:
The first step is to calculate capacitive reactance using Ohm's law. Since:

$$X_C = \frac{E}{I}$$

then:

$$X_C = \frac{220 \text{ V}}{3.2 \text{ A}}$$

$$X_C = 68.75 \ \Omega$$

$$C = \frac{1}{2\pi f X_C}$$

$$C = \frac{1}{(2 \ \pi \ 3.14159 \times 60 \times 68.75 \ \Omega)}$$

$$C = \frac{1}{25{,}918}$$

$$C = 3.86 \times 10^{-5} \text{ or } 38.6 \ \mu\text{F}$$

FIGURE 5–26   Capacitor circuit with .36 µF capacitor and 200 Hz supply.

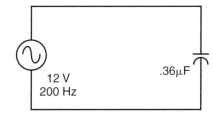

12 V
200 Hz

.36µF

Example

Take a look at Figure 5–26. What is the reactance of a .36 μF capacitor connected to a source that delivers 12 V at 200 Hz?

Solution:

$$X_C = \frac{1}{2\pi f C}$$

$$X_C = \frac{1}{(2 \times 3.14159 \times 200 \times .00000036)}$$

$$X_C = \frac{1}{.00045238}$$

$$X_C = 2210.5 \ \Omega$$

What is the current in this circuit?

$$I = \frac{E}{X_C}$$

$$I = \frac{12 \ \text{V}}{2210.5 \ \Omega}$$

$$I = 0.00543 \ \text{A, or } 5.4 \ \text{mA}$$

SUMMARY

Capacitors are devices used to store a charge and that stored charge has the potential to create current flow. This potential is the voltage stored in the capacitor. The ability to store a charge and become a voltage source gives it the characteristic that appears to oppose a change in circuit voltage. Capacitors are manufactured in many different ways but all have the same physical characteristics. The capacitor is made up of conducting plates separated by an insulator known as a dielectric. In a capacitive circuit, the current appears to flow through the capacitor because of the continuous increase and decrease of voltage and the continuous change in polarity in an AC circuit. The current flows back and forth in the circuit but does not flow through the capacitor. Electrons build up on one side of the capacitor and then on the other side as current alternates. Figure 5–27 illustrates current flow for half this cycle.

Because of the relationship of the apparent flow of current through the capacitor, and the voltage that appears across the capacitor, the capacitive current leads the capacitive voltage by 90°. This is an important point as the capacitive leading current is 180° out of phase—exactly opposite to the effects of an inductor which causes current to lag behind the voltage by 90°.

Time constants play an important part in timing circuits and in the basic understanding of how and why capacitors operate in various AC or DC circuits. The rate at which a capacitor charges or discharges is dependent upon the series resistance and the actual capacity of the cap. $R \times C$ allows us to calculate the seconds for one time constant, and a full charge or discharge is considered to be five time constants.

The current flow in the capacitive circuit is limited by capacitive reactance. Capacitive reactance is inversely proportional to the capacitance of the capacitor and the frequency of the line voltage source. Capacitive reactance is similar to inductive reactance, as it is measured in ohms and can be substituted for resistance in the Ohm's law formula. In a pure capacitive circuit, the current leads the applied voltage by 90°.

If the frequency of a circuit is increased, the capacitive reactance is decreased. If the capacitance is decreased, the capacitive reactance is increased.

FIGURE 5–27   One-half cycle current flow "through capacitor."

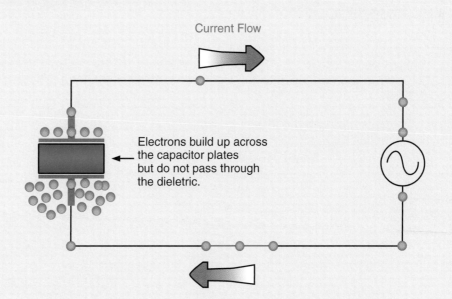

Current Flow

Electrons build up across the capacitor plates but do not pass through the dieletric.

REVIEW QUESTIONS

1. Explain what three physical factors must be present to have capacitance in a circuit.

2. What phenomenon causes a charge to be stored across the plates of a capacitor?

3. What are the units of measure for capacitance? Include the standard unit and 3 submultiples in your explanation.

4. What is the standard formula for farads when the physical characteristics of the capacitor are measured in the metric system?

5. Explain why no current flows through a capacitor but it appears that it does.

6. Provide a short explanation of the relationship of current and voltage in a capacitor.

7. Relate the current and voltage in an inductor to the voltage and current in a capacitor.

8. Explain the mnemonic "ELI the ICE man".

9. How are voltage ratings of a capacitor determined?

10. Explain the difference between dielectric strength and dielectric constant.

11. How many time constants does it take to charge a capacitor to 95% of its rated capacity?

12. Write two different formulas for determining X_C of a circuit when you know the frequency and the C rating of the capacitor.

13. Why does X_C decrease as the f increases?

14. In the capacitor formulas, the formula symbols have numerical measures. In what units are X_c, C, f, K, and d measured?

15. Draw a directional vector representing 50 ohms X_C on a diagram that uses resistance as the 0° reference.

PRACTICE PROBLEMS

1. A 20 μF capacitor operating in a circuit produces a voltage drop of 5 V at 1 kHz. What is the current?

2. What is the capacitance of a capacitor that has a reactance of 800 Ω at 10 kHz?

3. A 20 pF capacitor draws 10 mA at 95 V. What is the frequency?

4. An 80 μF capacitor is connected to a 240 V, 60 Hz supply. What is the current?

5. What is the charge in coulombs stored if a 50 μF capacitor is charged to 50 V?

6. Calculate the C for a capacitor with the following characteristics: 4 plates each .025 m², a dielectric K of 10, dielectric is .0002 meters thick.

7. Determine the voltage rating of the above capacitor if the $\dfrac{\text{volts}}{\text{cm}}$ is 50.

8. If a capacitor is in good operating condition, what would a test with an analog ohmmeter read when first connected?

9. If you double the size of the plate interaction, what happens to the capacitance?

10. If you reduce the dielectric thickness of a capacitor by ½, what happens to the voltage rating of the capacitor?

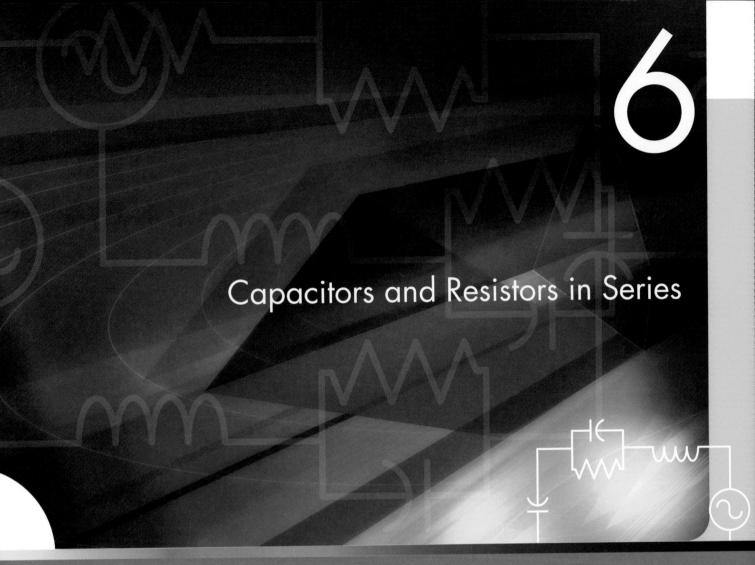

6

Capacitors and Resistors in Series

OVERVIEW

Capacitors are used in many AC circuits to react to the other circuit components. The ability to counteract the effects of the coil, or to balance a circuit with inductors, is one of the major uses of the capacitor. The fundamentals of the capacitor were presented in Chapter 5. In this chapter we will address the application of capacitors in various series circuits.

As you will see, many of the same principles used for the series equations for inductors also apply to the capacitor circuits. The connection of caps in a circuit in series has two major effects. The total capacitance of the circuit and the voltage rating of the circuit are both affected. Understanding the basic construction of the capacitor will make it easier to comprehend these variances from the inductor formulas.

The last section of the chapter includes a chart to help you decipher what will happen in general when changes are introduced to a circuit with resistors and capacitors in series.

OBJECTIVES

After completing this chapter, you should be able to:
- Solve for total capacitance in a series containing more than one capacitor
- Find total capacitive reactance in a series circuit containing more than one capacitor
- Solve for unknowns in circuits containing more than one capacitor in a series circuit when the values of other variables are specified
- Determine all the calculated values for oppositions, voltages, and currents in a series RC circuit
- Determine the dissipated power, the apparent power, and the reactive power of a series RC circuit
- Compute the power factor, the phase angle, and components of a power triangle
- Determine what happens to a series RC circuit as variables change

CAPACITORS IN SERIES

Connecting capacitors in series has the same effect as increasing (the actual distance between the plates does not increase) the distance between the plates, thereby reducing the total capacitance of the circuit. If you consider the circuit in Figure 6–1, with the same source voltage applied to each circuit, the circuit with two capacitors has twice as much dielectric thickness for the same applied circuit voltage. Using the formula:

$$C = \frac{.225 \text{ Ak}}{d}$$

where:
   C = the capacitance of a capacitor in farads
.225 = a constant for English units
   A = the effective plate area of one plate in square inches
   k = the dielectric constant of the dielectric material
   d = the distance between the plates measured in inches

FIGURE 6–1

A circuit with one capacitor compared to a circuit with two capacitors identical to the single capacitor circuit.

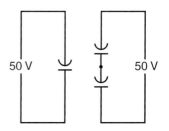

50 V   50 V

you can verify that if d increases, the C decreases. Each capacitor of the two in a series has more dielectric or less volts per mil than does the single capacitor circuit. The electrostatic field has the ability to hold less charge when the volts per mil thickness are less. In other words, the plates are effectively separated by a larger distance and the fields of the electrons have less influence.

By using the formula $C = \dfrac{Q}{V}$ we can again confirm the effects of adding capacitors in series. Take the capacitance of one capacitor and determine the Q (quantity of electrons) charge that can be accumulated on the plates. We manipulate the formula to $C \times V = Q$. With one-half the original voltage applied to the single series capacitor, the Q will be one-half the amount of a single capacitor in the same circuit. That one-half charge can be equated to current flow to the capacitor. Since the two capacitors are in series, that same amount of current will flow to the second capacitor. The electrons move off the bottom plate of C1 and move to C2. This capacitor will only hold the same charge, namely the same half charge compared to the original single capacitor. By looking at the current flow we can determine that only half the normal current will flow, thereby confirming that the total capacity of the circuit is only half the original capacitance.

Still another way to analyze the process is to use the formula for capacitive reactance: $X_C = \dfrac{1}{2 \pi f C}$, where C is in farads. If the circuit with one capacitor has an X_C of 10 ohms and the circuit with two capacitors has X_C of 10 ohms each, the total opposition for the two-cap circuit is twice as much as the one-cap circuit, like resistors. As you learned in the last chapter, we can determine the C of the capacitor if we know the X_C. Using the formula $C = \dfrac{1}{2 \pi f X_C}$ and inserting X_C into the denominator, you can deduce that the C is smaller when the X_C is larger.

This leads us to the formula for adding capacitors in series. As we add more capacitors in series the total capacitance of the circuit decreases, similar to what happens as we add resistors in parallel. When resistors are connected in parallel, there are several ways to calculate total resistance.

Examples:
- When all the resistors are of equal value, the total resistance is equal to the resistance of one of the resistors or branches divided by the number of resistors or branches:

$$R_T = \frac{R}{N}$$

- If two resistors of unequal resistance are in parallel, the formula for the total is:

$$R_T = \frac{R_1 \times R_2}{R_1 + R_2}$$

- If more than two unequal resistors are in parallel, the formula is:

$$\frac{1}{R_T} = \frac{1}{R_1} + \frac{1}{R_2} + \frac{1}{R_3} + \ldots$$

$$R_T = \frac{1}{\left(\frac{1}{R_1}\right) + \left(\frac{1}{R_2}\right) + \left(\frac{1}{R_3}\right) + \ldots}$$

SERIES CAPACITOR FORMULAS

The formulas for capacitance in a series circuit are as follows:
- When all the capacitors are of equal value, the total capacitance is equal to the capacitance of one of the capacitors or branches divided by the number of capacitors or branches:

$$C_T = \frac{C}{N}$$

- If two capacitors of unequal capacitance are in parallel, the formula for the total is:

$$C_T = \frac{C_1 \times C_2}{C_1 + C_2}$$

- If more than two unequal capacitors are in parallel, the formula is:

$$C_T = \frac{1}{\left(\frac{1}{C_1}\right) + \left(\frac{1}{C_2}\right) + \left(\frac{1}{C_3}\right) + \ldots}$$

where:
C = capacitance of one capacitor
N = number of equal capacitors connected in series and all the individual series capacitors are equal

Example

What is the total capacitance of three capacitors connected in series with the values of $C_1 = 20$ μF, $C_2 = 25$ μF, and $C_3 = 45$ μF?

Solution: All the *C* values are in microfarads so C_T is also in microfarads.

$$\frac{1}{C_T} = \frac{1}{C_1} + \frac{1}{C_2} + \frac{1}{C_3} \cdots$$

$$\frac{1}{C_T} = \frac{1}{20} + \frac{1}{25} + \frac{1}{45}$$

$$\frac{1}{C_T} = .05 + .04 + .022$$

$$\frac{1}{C_T} = .112$$

$$C_T = \frac{1}{.112}$$

$$C_T = 8.91 \ \mu F$$

Example

Two 50 μF capacitors are connected in series. Each capacitor is rated for 50 V.

1. Find the total circuit capacitance.

2. Find the total voltage that can be applied to the circuit.

Solution:

1. All the *C* values are in microfarads so *C* total is also in microfarads.

$$C_T = \frac{C_1 \times C_2}{C_1 + C_2}$$

$$C_T = \frac{50 \times 50}{50 + 50}$$

$$C_T = \frac{2500}{100}$$

$$C_T = 25 \text{ microfarads}$$

or

$$C_T = \frac{C_T}{N}$$

$$C_T = \frac{50}{2}$$

$$C_T = 25 \ \mu F$$

2. Because the dielectrics are each rated for 50 V (and the capacitors are in series) the total voltage of the circuit can be 100 V.

FieldNote!

When in the field, it is not always convenient to find the exact replacement for a given capacitor. If the circuit needs to be repaired quickly you may be able to find the right combination of capacitors to connect into the circuit. For instance if you need 20 μF and you have two 40 μF capacitors, you may connect them in series to yield 20 μF. Be sure to observe the voltage rating. If you have two 50 V capacitors they may be connected in series to yield 100 V rating but at a lower total capacity with two in series. You may install capacitors with a higher voltage rating than is required by the circuit but never a lower total voltage rating for your combination of capacitors than is required by the circuit voltage.

CAPACITIVE REACTANCE IN SERIES

As capacitors are added in series the total C decreases. However, the total X_C increases as it does with series resistors. Because each capacitor adds opposition to the circuit, the effect is cumulative. The formula for capacitive reactance is series is:

$$X_{CT} = X_{C1} + X_{C2} + X_{C3} \ldots$$

The total reactance can also be calculated by using the total values and applying the generic X_C formula:

$$X_{CT} = \frac{1}{2\pi f C_T}$$

Notice that as we added capacitors to a series circuit, the total C decreased. Using this lower total C, the effect on the X_{CT} in the formula is to cause it to increase; thereby verifying that total reactance increases as capacitors are added in series.

RESISTIVE CAPACITIVE SERIES CIRCUITS

All series circuits have two principles in common:
1. The current in all elements of the series circuit is equal to the total current.
2. The sum of the voltage drops across all circuit elements (including the sources) is equal to zero. Refer to Kirchoff's laws in your study of DC Theory.

A circuit that contains resistance and capacitance is called an RC circuit. You will find that the series RC circuit is similar to the RL circuit in that it uses the circuit current as the reference. The main difference is that the voltage drop across the capacitor lags the current by 90° (Figure 6–2).

FIGURE 6–2   Phase relationships of "perfect" RC and RL circuits with 90° shifted current.

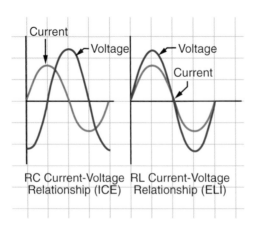

If a pure capacitive load (no resistance) is connected in an AC circuit, the voltage and current are 90° out of phase with each other. In a capacitive circuit, the current leads the voltage by 90° (Figure 6–3). Notice that (when evaluating series circuits) the voltage is always drawn as the reference at 0° on the horizontal.

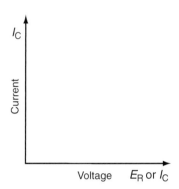

When a circuit that contains both resistance and capacitance is connected to an AC power supply, the voltage and current will be out of phase by some amount between 0 and 90°. The size of the phase angle will be determined by the ratio between resistance and capacitive reactance. The way we can determine what the circuit current will be is to construct the total opposition due to the resistance and the capacitive reactance.

RC series circuits are similar to resistive-inductive (RL) series circuits previously covered in Chapter 4. The formulas are basically the same with only minor modifications. Figure 6–4 shows a 20 Ω resistor in series with a capacitor having 30 Ω of capacitive reactance. The voltage supply is 220 V at 60 Hz. This circuit will be used for several of the examples in this chapter.

FIGURE 6–4   A Series RC circuit with 20 Ω resistor and 30 Ω of X_C capacitor.

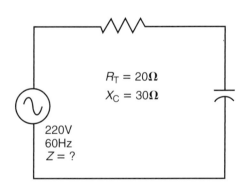

$R_T = 20\Omega$
$X_C = 30\Omega$

220V
60Hz
Z = ?

When we add the two oppositions, we need to remember that the ohmic values are both ohms but are two different types of opposition. Ohms of resistance are in phase with the applied voltage and are drawn as a vector at the 0° reference. The X_C causes an ohmic value at a 90° lead to the voltage as shown in Figure 6–5. The combination is *not* added arithmetically but instead added vectorially to get a resultant vector as the hypotenuse of the triangle. This resultant vector is again called the impedance as it was in the RL circuit.

TechTip!

The term "flux capacitor" may be familiar to many younger electricians. It was a term used in the movie *Back to the Future* to refer to the power production device on the vehicle that transported the main character back in time. The term is interesting as it combines the capacitor ideas of energy storage with the inductor ideas that use magnetic flux. As you know, the energy is stored in an electrostatic field in a capacitor and in an electromagnetic field in the inductors. Current discussion about electric cars is focused on how to store energy for use in the electric motor of these vehicles. This storage is now done with batteries but extensive research is being done to find better storage systems. One such new idea is to use capacitive storage. We must find a way to charge a capacitor with enough charge and then allow the charge to drive an electric motor. This discharge circuit resembles the RC discharge circuit described in Chapter 5.

FIGURE 6–5

X_C is drawn down from the horizontal representing that C causes a leading effect.

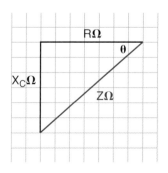

IMPEDANCE

Remember that total impedance is equal to the total opposition to current flow in the circuit. It is a combination of both resistance and capacitive reactance. Since this is a series circuit, the elements are vectorially added. Resistance and capacitive reactance are 90° out of phase with each other, and these form a right triangle with the impedance forming the hypotenuse (Figure 6–5). The formula for Z_T is:

$$Z_T = \sqrt{R^2 + X_C^2}$$

$$Z_T = \sqrt{20^2 + 30^2} = \sqrt{400 + 900}$$

$$Z_T = 36 \ \Omega$$

6.1 Adding Series Components

As we add series components, we still must follow the rules for series circuits. Opposition vectors that are at the same angle and direction can be added together. Therefore, as we add resistors and capacitors as depicted in Figure 6–6, we can create a circuit impedance triangle with all the oppositions in the circuit as in Figure 6–7.

FIGURE 6-6   Series circuit with multiple resistors and multiple capacitors.

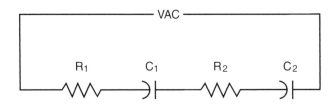

FIGURE 6-7   Vectors representing added Resistors and added capacitors.

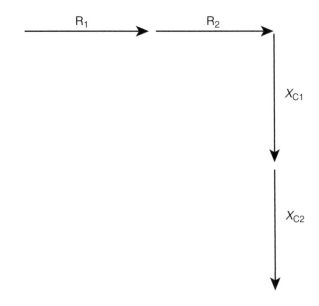

Example

Construct an impedance triangle using two resistors, each valued at 50 Ω, and a 26.5 μF capacitor at 60 Hz as shown in Figure 6–8.

Solution:
Total resistance is 100 Ω at 0° angle and X_C is 100 Ω at 90° angle (Figure 6–9).

$$Z_T = \sqrt{R^2 + X_C^2}$$

$$Z_T = \sqrt{100^2 + 100^2}$$

$$Z_T = \sqrt{10,000 + 10,000}$$

$$Z_T = 141.4 \ \Omega$$

FIGURE 6–8   Vector diagram with two 50Ω resistors and one 26.5 μF capacitor.

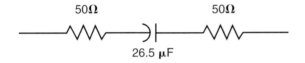

FIGURE 6–9   Impedance triangle showing calculation of total impedance.

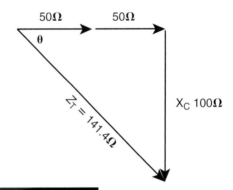

Example

Use the same components as above, but add a second 26.5 μF capacitor (Figure 6–10). Solve for the circuit impedance and for the circuit current if 200 V AC is applied.

Solution:
The impedance triangle appears as shown in Figure 6–11.

$$Z_T = \sqrt{R^2 + X_C^2} \ \text{using total R and total } X_C$$

$$Z_T = \sqrt{100^2 + 200^2}$$

$$Z_T = \sqrt{10,000 + 40,000}$$

$$Z_T = 223.6 \ \Omega$$

Using Z_T and 200 V AC applied:

$$I_T = \frac{E_T}{Z_T}$$

$$I_T = \frac{200 \ \text{V}}{223.6 \ \Omega}$$

$$I_T = .894 \ \text{A}$$

FIGURE 6–10 Series circuit with added component of 26.5μF capacitor.

FIGURE 6–11 Impedance triangle with all four component vectors.

50Ω   50Ω

26.5 μF   26.5 μF

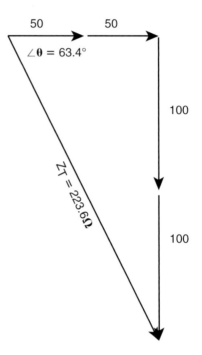

50   50

$\angle \theta = 63.4°$

100

$Z_T = 223.6\Omega$

100

CALCULATING VOLTAGE DROPS FOR SERIES RC COMPONENTS

Since the current is the same value at any point in a series circuit, the voltage drop across each capacitor can be computed using the capacitive reactance of each capacitor and the current flow of the circuit. Unlike the calculations for inductors that needed a Z triangle for each coil, capacitors have so little resistance compared to the X_C that the R is ignored. Therefore, we just use the circuit current that appears to flow through the capacitor multiplied by the component opposition (X_C) to find the voltage drop across the capacitor.

Example

Calculate the voltage drops for each of the capacitors in Figure 6–12 (see page 130).

Solution:
Step 1
Find the capacitive reactance of each capacitor:

$$X_C = \frac{1}{2\pi fC}$$

$$X_C = \frac{1}{(377)\,\pi fC} \quad \text{Note: 377 is only correct when the } f \text{ is 60 Hz.}$$

$$X_{C1} = 132.7 \ \Omega$$

$$X_{C2} = \frac{1}{2\pi f C}$$

$$X_{C2} = \frac{1}{(377)\,25\mu F}$$

$$X_{C2} = 106.1\ \Omega$$

$$X_{C3} = \frac{1}{2\pi f C}$$

$$X_{C3} = \frac{1}{(377)\,45\mu F}$$

$$X_{C3} = 58.9\ \Omega$$

Step 2

Now find the total capacitive reactance:

$$X_{CT} = X_{C1} + X_{C2} + X_{C3}$$

$$X_{CT} = 132.7 + 106.1 + 58.9$$

$$X_{CT} = 297.7\ \Omega$$

Step 3

Now find the current in the circuit:

$$I_T = \frac{E_T}{X_{CT}}$$

$$I_T = \frac{480V}{297.7\ \Omega}$$

$$I_T = 1.61\ A$$

Step 4

Now find the voltage drop across each capacitor:

$$E_C = I_C \times X_C\ (\text{for any one capacitor})$$

$$E_{C1} = 1.61\ A \times 132.7\ \Omega$$

$$E_{C1} = 213.6\ V$$

$$E_{C2} = 1.61\ A \times 106.1\ \Omega$$

$$E_{C2} = 170.8\ V$$

$$E_{C3} = 1.61\ A \times 58.9\ \Omega$$

$$E_{C3} = 94.8\ V$$

Remember that to check this, you can add the voltage drops together. The sum of the individual voltage drops should be equal to the total voltage from the supply, because they are all voltages at 90° lagging behind the current reference.

$$E_T = E_{C1} + E_{C2} + E_{C3}$$

$$E_T = 213.6\ V + 170.8\ V + 94.8\ V$$

$$E_T = 479.2\ V$$

The .8 volts of the final answer, 479.2 V, is caused by math round-off errors throughout the calculations.

FIGURE 6–12 A 3 capacitor series circuit used to calculate voltage drop on each component.

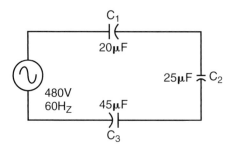

IMPEDANCE AND VOLTAGE TRIANGLES FOR RC CIRCUITS

Refer back to Figure 6–10 with two resistors and two capacitors in series. As we calculated earlier, the impedance was at 223.6 Ω and the circuit current was at .894 A when 200 V AC was applied. This created a Z triangle that appears in Figure 6–11. The angle to the horizontal current reference was 63.4° leading current or can also be stated as lagging voltage. This impedance triangle is congruent to the voltage triangle of Figure 6–11. We can calculate the voltage drop across each component by using the total current of the series circuit as a constant. This current is multiplied by the opposition of each component and the voltage appears as the vector with the same characteristics as the opposition. The resistive components are drawn at 0° angles and the voltage across the capacitors is drawn at 90° lagging to the horizontal. Now we can vectorially add all the vectors to verify the circuit total voltage.

Example

$$E_{R1} = I_{R1} \times R_1$$

$$E_{R1} = .894 \times 50 \ \Omega$$

$$E_{R1} = 44.7 \ \text{V}$$

$$E_{R2} = .894 \times 50 \ \Omega$$

$$E_{R2} = 44.7 \ \text{V}$$

$$E_{C1} = .894 \times 100 \ \Omega \ \text{of} \ X_C$$

$$E_{C1} = 89.4 \ \text{V}$$

$$E_{C2} = .894 \ \text{A} \times 100 \ \Omega \ \text{of} \ X_C$$

$$E_{C2} = 89.4 \ \text{V}$$

Total resistive voltage drop is approximately 89.4 V and total capacitive voltage drop is approximately 178.8 V. Adding the vectors together and using either trigonometry or the Pythagorean theorem, the resultant hypotenuse is 199.9 V or very nearly 200 V, thus confirming that we started with 200 V AC. We will not calculate exactly 200 V, as we rounded some of our calculations.

We now have matching (similar or congruent) impedance (Z) triangles and voltage (E) triangles vectors. We can again calculate the angle between the reference line current on the horizontal and the angle of the resultant voltage, or the circuit total applied voltage. We use the vectors at 90° angles, 178.4 V capacitive and 89.4 V resistive. Through trigonometry we use 178.7 V (the opposite side) divided by 89.4 V (the adjacent side) and using the arc which has a tangent of 2, we calculate the angle of leading current at 63.4°, just the same as the Z triangle.

POWER

As with the inductive circuits, the power is categorized by the relationship of the circuit current to the circuit voltage. The zero axis reference is the in-phase current and voltage. Again, there are three components to the RC power triangle. True power (watts) is still calculated by the line current and the resistance of a resistor using I^2R, or by using $I_T \times E_R$. The volt-amps-reactive power (VARs) are calculated by using $I^2 \times X_C$ or $I_T \times E_C$. With these two components we can build another right triangle that is again congruent to the Z and the E triangles (Figure 6–11). The resultant side, the hypotenuse, is the

TechTip!

By adding the individual voltages arithmetically we get 89.4 + 178.8 V to yield 268.2 V; more than the 200 V with which we started. This practice of adding voltages together arithmetically to total more than originally applied, often confuses even experienced electricians who have forgotten the effects of capacitor and resistor individual voltages. We get so used to determining what is to happen in a series circuit that we mistakenly apply DC rules to AC circuits.

FIGURE 6–13 A power triangle with appropriate formulas on each side.

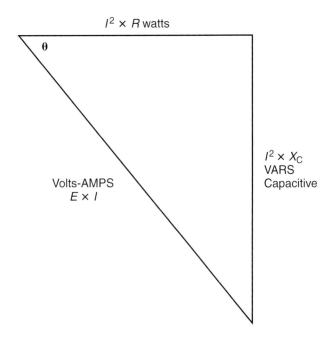

apparent power of the circuit. When we multiply line current times line voltage we apparently get power. Again, this is apparent power, which is not the same as true or dissipated power. Using the same circuit as Figure 6–10 with two 50 Ω resistors and two 26.5 μF capacitors at 200 V AC and 60 Hz, we assume the power is .894 A × 200 V = 178.8 volt-amps (VA). True power is the power that is dissipated where current and voltage are in phase, namely the horizontal portion of the power triangle. In this case the true power is .894 A × 89.4 V ≈ 80 W (rounded off). The reactive power is .894 × 178.8 V ≈ 160 VARs (rounded off). The power triangle appears in Figure 6-14 with the same angle of 63.4° to the horizontal proven by trigonometry.

FIGURE 6–14   Power vectors for the circuit depicted in Figure 6–12.

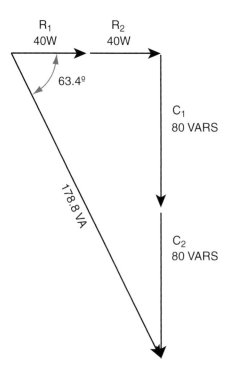

The power factor of this circuit can also be calculated the same as the RL circuits. The cosine of the phase angle × 100 equals the percent power factor or by using the hypotenuse and the adjacent side to the phase angle we find that $\left(\dfrac{\text{watts}}{\text{volt-amps}}\right) \times 100 = \%$ power factor (*PF*). For the circuit above, the watts are 80 W and the VAs are approximately 179. Therefore:

$$\left(\frac{80\ \text{W}}{179\ \text{VA}}\right) \times 100 = 44.7\%\ PF$$

To corroborate with trigonometry, the cosine of 63.4° is .447 × 100 = 44.7% *PF*. This means that 44.7% of the power delivered to the circuit in the form of volts and amps is actually used or consumed as watts. The remaining energy is stored in the form of electrostatic fields in the capacitor. In this case the stored power is caused by a leading current (ICE) and the power factor is referred to as a leading power factor. This is an important point because the inductors caused a lagging power factor due to a lagging current.

6.2 Angle Theta

The power factor of the circuit is the cosine of the phase angle. Since the power factor turned out to be 0.447, or 44.7 %, angle theta (θ) can be calculated:

$$\cos \theta = PF$$

$$\cos \theta = .447$$

$$\text{arc cos } .447 = \theta$$

$$\theta = 63.4°$$

Therefore, if we know the power factor for either leading power factor or lagging power factor we can find the actual angle that the current leads or lags the voltage. Likewise if we know the angle, we can use the cosine function to find the power factor.

Example

Find the power factor of the following circuit. A single 120-Ω resistor is in series with two 40-μF capacitors and the circuit is operating at 50 Hz and 200 V AC (Figure 6–15).

FIGURE 6–15   Circuit with two capacitors and a resistor operating at 50 Hz.

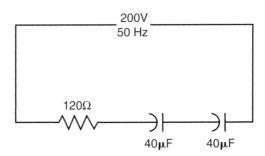

Solution

Find the impedance triangle by determining the X_C of the capacitors. Knowing that two 40 μF capacitors in series are the same as one 20 μF, capacitor we use the formula

$$X_C = \frac{1}{2\pi fC} = \frac{1}{2 \times 3.14 \times 50 \times 20 \ \mu F}$$

$$X_C = 159.2 \ \Omega$$

$$R = 120 \ \Omega$$

$$Z = \sqrt{R^2 + X_C^2}$$

$$Z = \sqrt{120^2 + 159.2^2} \quad \text{(Figure 6–16)}$$

$$Z = 199.4 \ \Omega$$

FIGURE 6–16 Impedance triangle for circuit in Figure 6-15.

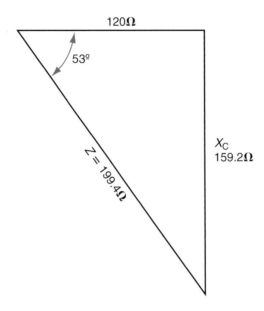

The power factor is the cosine of the angle between the hypotenuse and the resistance. That angle is $\approx 53°$

$$\text{Cos } 53° = .60$$

$$PF = .60 \times 100 = 60\%$$

We can go directly to the power triangle to confirm this figure by first finding the line current.

$$I_T = \frac{E_T}{Z_T}$$

$$I_T = \frac{200 \text{ V}}{199.4 \text{ } \Omega}$$

$$I_T \approx 1 \text{ A}$$

Then the power triangle looks like Figure 6–17.

Apparent power is $VA = V \times I_T$

$$VA = 200 \text{ V} \times 1 \text{ A}$$

$$VA = 200 \text{ } VA$$

$$W = I^2 \times R$$

$$W = 1^2 \times 120 \text{ } \Omega$$

$$W = 120 \text{ } W$$

FIGURE 6–17 Power triangle for circuit in Figure 6–15.

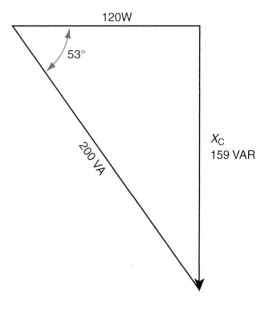

Then the power factor is: $\dfrac{W}{VA}$

$$\% \, PF = \dfrac{W}{VA} \times 100$$

$$\% \, PF = \dfrac{120}{200} \times 100$$

$$\% \, PF = 60$$

The angle of lead is arc cos .60 = 53.1°. This is approximately the same as the angle found in the Z triangle. It should be identical but is not because we rounded figures to make calculation procedures more clear.

THE EFFECTS OF CHANGING VARIABLES ON RC CIRCUITS

The effects of frequency on a series RC circuit can be analyzed by determining what happens to the X_C of a capacitor. This effect can then be used to determine what happens to the impedance triangle. The Z triangle affects the circuit current and the voltage dropped on each of the series components. This in turn affects the total power triangle. Let us assume a 100 µF capacitor that has 60 Hz applied. The circuit has a 25 Ω resistor in series (Figure 6–18).

If we decrease the applied frequency to 30 Hz, the X_C goes from 26.5 Ω to 53 Ω. The shape of the impedance triangle changes as well (see Figure 6–19). The circuit impedance goes up and the line current goes down. The voltage triangles change to be consistent with the Z triangles as in

FIGURE 6-18   A circuit containing a simple resistor and capacitor in series and the Z triangle.

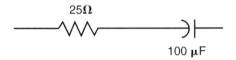

25Ω

100 μF

Figure 6–19. The voltage across the capacitor increases because the X_C increases at a higher proportion than the current decreased. The power triangle also changes to reflect the Z and E triangles. The power factor changes to become a lower percent power factor to coincide with the larger leading current angle as the circuit appears more capacitive. The watts decrease because the line current decreased, and the resistance does not change.

FIGURE 6-19   The shape of the impedance triangle changes as the frequency goes to 30Hz.

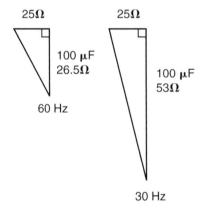

25Ω   25Ω

100 μF
26.5Ω   100 μF
                        53Ω
60 Hz

30 Hz

Refer to Table 6–1 for effects of frequency, capacitance, or resistance on RC series circuits when changing one variable at a time.

As you can see from Table 6–1, many of the circuit conditions that may change have effects on the entire circuit. The results of a change in the circuit parameters may take some effort to determine because of the right triangle vector additions. A linear change in one variable does not always create a linear change in the results.

TABLE 6-1 Use this table to compare what will happen if you change one variable at a time in a series RC circuit. The top line identifies the calculated value, and the left column indicates which was the variable changes. Use the chart to predict the effects of different calculations as the variables are changed one at a time. Use the impedance, voltage, and power triangles to verify trends.

Variable	X_C	Impedance	Line Current	Capacitor Voltage	Phase Angle	PF	Watts
Freq Inc	Dec	Dec	Inc	Dec	Dec	Inc	Inc
Freq Dec	Inc	Inc	Dec	Inc	Inc	Dec	Dec
Cap Inc	Dec	Dec	Inc	Dec	Dec	Inc	Inc
Cap Dec	Inc	Inc	Dec	Inc	Inc	Dec	Dec
Resis Inc	Same	Inc	Dec	Dec	Dec	Inc	Dec
Resis Dec	Same	Dec	Inc	Inc	Inc	Dec	Inc

Inc = increase; Dec = decrease

SUMMARY

In a pure capacitive circuit, the voltage and current are 90° out of phase with each other. In a pure resistive circuit, the voltage and current are in phase with each other. In a circuit containing both resistance and capacitance, the voltage and current will be out of phase with each other by some angle between 0 and 90°. The amount of phase angle difference between voltage and current in a series RC circuit is equal to the angle whose tangent is the ratio of resistance to capacitive reactance.

To find the total capacitance of a series of capacitors, the formulas look like the formulas for parallel resistors with the total C decreasing as more capacitors are added in series. This effect is the result of adding more dielectric thickness to the overall circuit voltage. Another effect of adding more dielectric material is that the overall voltage rating of the series of capacitors is equal to the arithmetic sum of the individual capacitor's voltage rating.

When we add a resistor in series with the capacitor, we must construct an impedance triangle to determine the total opposition in the circuit. The capacitor's X_C is always drawn at 90° to the horizontal and resistance is always drawn at 0°. The resultant side (the hypotenuse) is the total Z at an angle θ to the horizontal. This is the angle that the current leads the voltage of the circuit. This same angle will determine the power factor of the circuit by taking the cosine of angle θ. The same angle (θ) is replicated in the voltage triangle vector for the circuit and for the power triangle for the circuit. These two triangles are congruent to the Z triangle. The E triangle vector uses the circuit current and the R and X_C to determine the volt drops across each component. The hypotenuse should again equal the applied line voltage. The power triangle is made of vectors for watts, VA, and VARs of capacitive power.

Once the circuit components and triangles are calculated, it makes it easier to determine what will happen to the circuit parameters as various variables change.

REVIEW QUESTIONS

1. Why does the total capacitance of a circuit decrease as more capacitors are added in series?

2. Write a formula for finding C_T if you have two capacitors of the same value in series.

3. Find the C1 of a circuit if C2 is 50 μF and C3 is 75 μF and the total capacitance is 23 μF.

4. Calculate the C of a capacitor if the voltage across the capacitor is 50 V, the circuit current is 2 A, and the circuit frequency is 60 Hz.

5. Explain what happens to the X_C of a capacitor as the frequency of a circuit increases.

6. As multiple resistors are added to multiple capacitors in series, how do you calculate the effects on the circuit current?

7. Explain what happens to the power factor by adding resistors to a series RC circuit.

8. Explain how the angle of leading current changes as capacitors are added in series.

9. Write the formula for calculating the total capacitive reactance of a circuit as capacitors are added in series.

10. Find the VA for a circuit with a power factor of 80% and a watt measurement of 200 W.

PRACTICE PROBLEMS

1. Current _____ voltage in an RC circuit.
2. The reference vector in a series circuit is generally the _____ vector on the horizontal axis.
3. Draw and label an impedance (phasor diagram) triangle for a series RC circuit.
4. In Figure 6–20, if $E_C = 80$ V, what is the voltage across the resistor?
5. What is the X_C in Problem 4?

6. What is I_T?
7. What is the value of the resistor in Problem 4?
8. Find the impedance of the circuit of Problem 4. Prove it.
9. Find *PF* in Figure 6–20
10. A circuit has $E_T = 15$ V at 2 kHz with a resistor (3 kΩ) and a capacitor in series. If 4 mA flow, what is the value of the capacitor?

FIGURE 6–20   Circuit for Practice Problems 4–9.

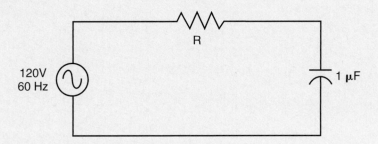

7

RLC Series Circuits

OVERVIEW

You now know that inductance causes current in the circuit to lag behind the applied voltage and that capacitance causes the current to lead the applied voltage. The voltage across an inductor leads the current by 90°, and the voltage across the capacitor lags the current by 90°. Remember that current in a series circuit is the same at all points and therefore can be used as a reference to analyze the circuit. As stated before, using current for the reference is for learning purposes. Most technical proofs use vector diagrams with voltage as the reference. We need to determine the effects on the series circuits when both inductance and capacitance is connected. Although the steps to solve the circuit calculations are the same as the steps for either RL or RC, the combination creates new procedures. With the real world of inductors that actually have R and X_L, the process for finding circuit quantities is a little more complicated. The need to keep accurate account of drawings and assigned values becomes more apparent. With the added dimension of changing line frequency of the circuit, the dynamics of changing X_L and X_C become more important.

OBJECTIVES

After completing this chapter, you should be able to:
- Determine expected circuit values based on calculations
- Draw and describe vector diagrams representing circuit impedance
- Determine voltage drops across all components and verify against circuit totals
- Describe the behavior of an RLC circuit regarding current and voltage relationships
- Calculate expected power values for each component and for circuit totals in a series RLC circuit
- Describe and analyze what happens in a series RLC circuit at, or near, resonance
- Find all values in a series circuit with R, C, and coil Z

RLC SERIES CIRCUITS

7.1 Circuit Calculations

The methods used to calculate what will occur as you build series circuits with resistive, inductive, and capacitive components are combinations of the last several chapters where we treated the circuits separately. When an AC circuit contains the elements of resistance, capacitance, and inductance connected in series, the current is the same through all parts, but the voltage drops across the components are out of phase with each other. This is a characteristic of a series circuit.

The sequence of finding the total circuit effects are the same as previously described. The total impedance of the circuit is determined by the resistance at 0° to the horizontal, the X_L at 90° lagging, and the X_C at 90° leading vectors. From these principles, you can conclude that in a series inductive-capacitive (LC) circuit, X_L and X_C are 180° out of phase with each other. Therefore, the value of one will subtract from the other (vectors in opposite directions) and the remainder will determine if the circuit is either inductive or capacitive. To find the effects on the total impedance of the circuit with reactive components of opposition use the formula:

$$Z_T = \sqrt{R^2 + (X_L - X_C)^2}$$

The value of the difference between X_L and X_C is simply a matter of subtracting vectors exactly opposite of each other. Just keep track of which reactive component is larger as you square the difference. We are creating a total effect-impedance triangle and we need to know which opposition has a greater effect (Figures 7–1 and 7–2).

FIGURE 7–1 A series or circuits that contains resistance, inductance, and capacitance.

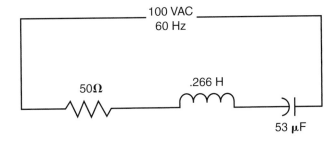

Example

Using the circuit in Figure 7–1 find the oppositions and the current based on formulas learned in previous chapters.

Given:

- Resistor (R_1) = 50 Ω

  Inductor (L_1) = .266 H (no resistance to this example coil)

$$X_L = (2\pi f)\, L$$

$$X_L = (6.28f)\, .266$$

$$X_L = 100\ \Omega$$

- Capacitor (C_1) = 53 μF

$$X_C = \frac{1}{2\pi f C}$$

$$X_C = \frac{1}{(6.28f) \times .000053}$$

$$X_C = 50\ \Omega$$

- $Z_T = \sqrt{R^2 + (X_L - X_C)^2}$

$$Z_T = \sqrt{50^2 + (100 - 50)^2}$$

$$Z_T = 70.7\ \Omega$$

- $I_T = \dfrac{E_T}{Z_T}$

$$I_T = \frac{100\ \text{V}}{70.7\ \Omega}$$

$$I_T = 1.414\ \text{A}$$

See Figure 7–2 for total impedance resultant of 70.7 Ω.

FIGURE 7-2   Impedance vectors and resultant values.

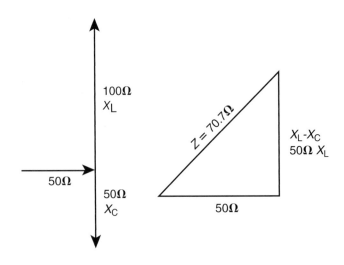

As you will note in Figure 7–2, the angle theta (θ) is 45° and the effect of the circuit as it appears in the vector diagram is more inductive because of the effects of the inductive reactance. Therefore the circuit current will react like the current in an inductive circuit and lag the voltage by 45°. This same angle will be replicated in the voltage vector and the power triangle as it was in the previous chapters. The ratio of total reactance to resistance will determine how much the applied voltage will lag or lead the circuit current. If the circuit is more inductive than capacitive, the current will lag the applied voltage, and the power factor will be a lagging power factor. If the circuit is more capacitive than inductive, the current will lead the voltage, and the power factor will be a leading power factor.

Since inductive reactance and capacitive reactance are 180° out of phase with each other, they subtract from each other in an AC circuit. This subtraction may mean a total reactance that could be zero ohms. This means that the current may be very high, depending on the value of R. Remember that $\dfrac{E_T}{Z_T}$ equals I_T.

7.2 Voltage Drop

The voltage drop across the resistor will be in phase with the current. The voltage drop across the inductor will lead the current by 90°, and the voltage drop across the capacitor will lag the current by 90°. The three sine wave diagrams in Figure 7–3 show this.

FieldNote!

When referring to the power of a circuit, the terminology for leading and lagging power factor refers to the relationship of the circuit current to the circuit voltage. If the total circuit current leads the voltage, then the circuit is said to have a "leading power factor." If the total circuit current lags the circuit voltage, then the circuit would have a "lagging power factor."

FIGURE 7–3   Voltage and current relationship in resistance, pure inductors and capacitors.

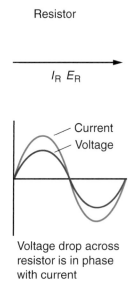

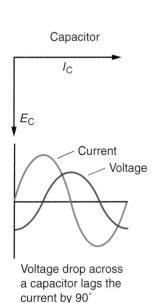

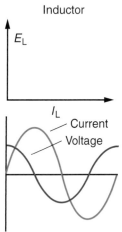

Voltage drop across resistor is in phase with current

Voltage drop across a capacitor lags the current by 90°

Voltage drop across the inductor leads the current by 90°

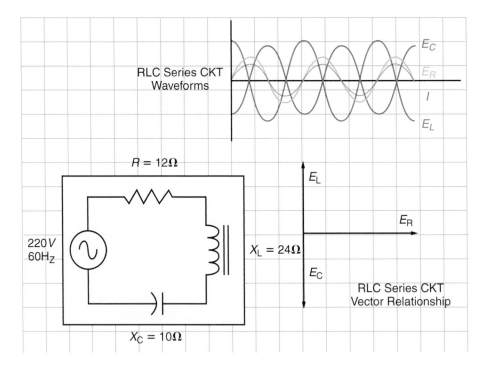

FIGURE 7–4   Voltage relationships that occur in a Series RLC circuit shown with circuit current.

Look at Figure 7–4. The sine waves, shown at the top, represent total current and the voltage across each element. Notice that E_R is in phase with the current, while E_C and E_L are 90° lagging or leading, respectively. This means that the inductive voltage and the capacitive voltage are 180° out of phase with each other.

The vectors representing those component values are also 180° out of phase, as seen in the vector diagram (Figure 7–5). These relationships are critical to understanding how the series RLC circuit components interact and perform.

FIGURE 7–5   Vector diagrams to match voltage relationships.

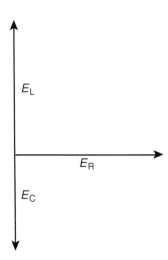

Also, when Ohm's law is used on the circuit values, the sum of the voltage drops developed across these components can exceed the applied voltage. Using the same components as in the previous example, we can construct the congruent voltage triangle vector diagram to see how much voltage will appear across each component. We must use the original values of oppositions to determine what the actual value of voltage is, even though the net effect in the circuit will again be the vector addition of different direction vectors (Figure 7–6).

FIGURE 7–6   Vector diagrams that illustrate circuit component values.

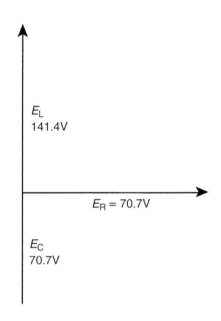

E_L
141.4V

$E_R = 70.7$V

E_C
70.7V

Example

Given:

$$R_1 = 50 \ \Omega$$

$$X_{L1} = 100 \ \Omega$$

$$X_{C1} = 50$$

$$I_T = 1.414 \ \text{A}$$

- $E_{R1} = I_T \times R_1$

$$E_{R1} = 1.414 \ \text{A} \times 50 \ \Omega$$

$$E_{R1} = 70.7 \ \text{V}$$

- $E_{L1} = I_T \times X_{L1}$

$$E_{L1} = 1.414 \ \text{A} \times 100 \ \Omega$$

$$E_{L1} = 141.4 \ \text{V}$$

- $E_{C1} = I_T \times X_{C1}$

$$E_{C1} = 1.414 \text{ A} \times 50 \text{ }\Omega$$

$$E_{C1} = 70.7 \text{ V}$$

- Angle 45° lagging current
- $E_T = \sqrt{E_R^2 + (E_L - E_C)^2}$

$$E_T = \sqrt{70.7^2 + (141.4 - 70.7)^2}$$

$$E_T \approx 100 \text{ V (rounded)}$$

The current flow is the same throughout the series circuit, so 1.414 A will flow through all the components. With this current flowing in each component in the circuit it would appear that the voltages all add up to more than the applied. As you now know, the voltage vectors have the same rules as the opposition vectors. As such the effect on the circuit of the two reactive voltage vectors is to subtract, and the new total voltage triangle appears as shown in Figure 7–7. We now prove that the final triangle voltage vector can be constructed and we can use either trigonometry or the Pythagorean theorem to prove the resultant hypotenuse is the original line voltage of 100 V.

FIGURE 7–7   Voltage triangle represents actual voltage measurements and total.

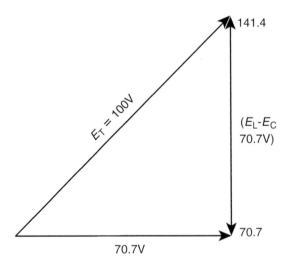

7.3 Power in RLC Series Circuits

Just as we have seen in series *RL* or series *RC* circuits, the power triangle follows the same patterns as the *Z* triangle and the *E* triangle vectors. (See Figure 7–8 for a power triangle.) The watts, or true power, is found on the horizontal line of the power triangle and can be calculated by current times the voltage across the resistive components or by the current squared times the resistance. The volts-amps-reactive (VARs) is found by the circuit current times the voltage across the capacitive components or the circuit current times the inductive voltages.

FIGURE 7–8   Power triangle for RCL series circuit.

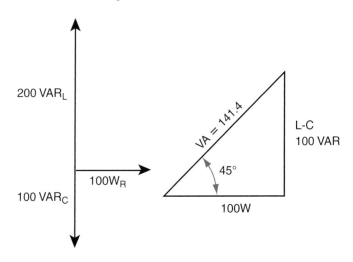

FieldNote!

When referring to VARs in the circuit, several terms are used for indicating that the volt-amps-reactive are capacitive. Sometimes they are called CVARs; they can also be written as -VARs or shown as VAR_C. These notations are used by different electrical personnel, but still denote that the reactive power is capacitive in nature.

These values can also be found by the current squared times the capacitive reactance or times the inductive reactance. Finally, after subtracting the VARs capacitive and the VARs inductive, a final VARs component will remain either more capacitive or more inductive. The two VARs can completely cancel each other to yield no reactive power in the final circuit and only watts remain. In that case the horizontal line is all that remains as watts and the power factor are 100%.

Example

Given:

$R_1 = 50 \ \Omega$

$X_{L1} = 100 \ \Omega$

$X_{C1} = 50 \ \Omega$

$I_T = 1.414 \ A$

- Watts

  dissipated by $R_1 = I_T \times E_R$

  $W_{R1} = 1.414 \ A \times 70.7 \ V$

  $W_{R1} \approx 100 \ W$ (rounded)

- VAR

  reactive power for $L_1 = I_T \times E_L$

  $VAR_{L1} = 1.414 \ A \times 141.4 \ V$

  $VAR_{L1} \approx 200$ VARs inductive (rounded)

- VAR

  reactive power for $C_1 = I_T \times E_C$

  $VAR_{C1} = 1.414 \ A \times 70.7 \ V$

  $VAR_{C1} \approx 100$ VARs capacitive (rounded)

- $VA_{\text{resultant apparent power}} = \sqrt{W^2 + (VAR_{\text{L1}} - VAR_{\text{C1}})^2}$

$$VA = \sqrt{100^2 + (200 - 100)^2}$$

$$VA = 141.4\ VA$$

or

$$VA = I_\text{T} \times E_\text{T}$$

$$VA = 1.414 \times 100$$

$$VA = 141.4\ VA\ \text{reactive power}$$

- Phase angle between the hypotenuse and the horizontal is 45°
- % Power factor $(PF) = \cos 45°$
- $\%\ PF = 70.7\%$

$$\%\ PF = \left(\frac{W}{VA}\right) \times 100$$

$$\%\ PF = \left(\frac{100}{141.4}\right) \times 100$$

$$\%\ PF = 70.7\%$$

7.4 Power Factor

The power factor is calculated by dividing the true power of the circuit by the apparent power. The answer is multiplied by 100 to turn the decimal into a percentage; the percent power factor is also the cosine of the phase angle times 100. Cosine is calculated by the adjacent side divided by the hypotenuse or watts divided by VA. In this case we expect the power factor to be 70.7%, because the angle is the same in all the triangles.

![FieldNote!]

FieldNote!

Power factor meters, watt meters, and VAR meters are available to measure these components of a power system. Power factor meters are used to measure the % PF and whether the factor is for a leading power factor or a lagging power factor. Watt meters use current sensors and voltage sensors to determine the "in phase" power. VARs can also be measured by VAR meters. These meters shift the voltage to match the out of phase current and then directly measure volts-amps-reactive (VARs). See Figure 7–9 for example of a meter.

FIGURE 7–9   Meter can measure Watts, VARs, and PF with voltage probes and current sensors.

FREQUENCY EFFECTS AND SERIES RESONANCE

When an inductor and capacitor are connected in series, there is a frequency at which the inductive reactance and capacitive reactance will become equal. This is because, as frequency increases, inductive reactance increases and capacitive reactance decreases.

The point at which two reactance values become equal is called **resonance**. At resonance, all reactive components are canceled and only the resistance of the circuit remains. This of course reduces the impedance of the circuit, which in turn raises the circuit current. The phase angle goes to 0° and the PF goes to 100%. The origin of the resonant frequency formula is to equate the X_C equal to the X_L. Resonant circuits can be used to provide increases of current and voltage at the resonant frequency.

Since $X_L = 2\pi fL$ and $X_C = \dfrac{1}{2\pi fC}$, then we know that when $2\pi fL = \dfrac{1}{2\pi fC}$ the two reactance values are equal. For any series circuit that has inductance and capacitance, we can insert the values of total L and total C to find which frequency will cause resonance. The formula used to determine the resonant frequency when the values of L and C are known is as follows:

$$X_L = X_C$$

$$2\pi fL = \frac{1}{2\pi fC}$$

$$f^2 = \frac{1}{(2\pi C)(2\pi L)}$$

$$\sqrt{f^2} = \sqrt{\frac{1}{(2\pi C)(2\pi L)}}$$

$$f_R = \frac{1}{2\pi\sqrt{LC}}$$

where:
f_R = frequency at resonance
L = inductance in henries
C = capacitance in farads

Example

Find the resonant frequency for a series circuit with an inductance of .0154 H and a capacitance of 1.61 µF.

Solution:

$$f_R = \frac{1}{2\pi\sqrt{LC}}$$

$$f_R = \frac{1}{2 \times 3.14 \times \sqrt{(0.0154 \times 0.00000161)}}$$

$$f_R = \frac{1}{6.28 \times (0.00016)}$$

$$f_R = 995 \text{ Hz}$$

This circuit will reach resonance at 995 Hz when both the inductor and the capacitor produce equal reactances. At this frequency the two reactances are moving in opposite directions, and since they are equal, they cancel each other out.

Resonance, or Resonant frequency
The value in hertz at which inductive reactance and capacitive reactance are equal (in resonance) in a circuit. At series resonant frequency, the circuit impedance is at a minimum and the circuit current is at a maximum.

TechTip!

One of the characteristics of the series resonant circuit is that when the circuit has resonant frequency applied, the circuit impedance goes to minimum, the circuit current goes to maximum, and the circuit component voltage drops will increase. This characteristic is often used in sensor technology in automated systems where sensors are used to detect some type of material within its sensing range. When a "target item" is within sensing range, the frequency of the circuit changes; going to resonance. This change in frequency causes the voltages within the sensor circuit to change, triggering an indication of a presence of the intended target.

Another use for resonant series circuits is to either block or to pass certain frequencies in electronic circuits. We can use the idea that at resonance, component voltages in the series circuit will increase. We can use this increased voltage to pass on a signal to another section of electronics or we can use the same effect to stop the signal from continuing. This is described in more detail in the band pass or band stop filters chapter.

When a circuit is not at resonance, current flow is limited by the combination of inductive reactance and capacitive reactance. When frequency is lowered, the inductive reactance decreases; however, since the capacitive reactance increases, the total impedance increases. If the frequency is increased above the resonant frequency, the inductive reactance increases, the capacitive reactance decreases, and the total impedance increases. In either case, there is more impedance than when both reactances cancel each other.

Take a look at Figure 7–10 to see the effects on current at resonance and at either higher or lower than resonant frequency. When the circuit reaches resonance (the frequency at which both L and C reactances are equal), the current will suddenly increase because the only current-limiting factor will be wire resistance. The effects on the circuit voltages can be predicted. At resonance the circuit current is at maximum. That current flows through each component and the voltage across each component rises. The true power of the circuit goes up because the current peaks at resonant frequency and the resistance values stay constant. At either side of resonance, the X_C and X_L are changing. With a lower than resonant frequency the circuit appears more capacitive, at a higher than resonant frequency the circuit appears more inductive. The phase angle and the power factor change rapidly around the resonant frequency point.

FIGURE 7–10   Chart shows circuit current at resonant frequency for Series RCL circuit. Impedance goes to a minimum at resonant frequency.

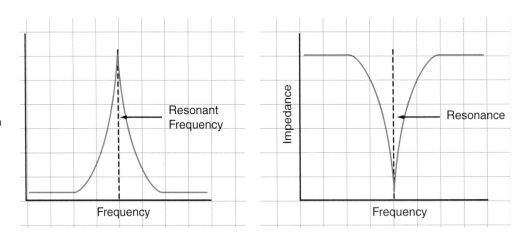

SERIES CIRCUIT WITH COIL RESISTANCE

Two resistors, two capacitors, and a coil with a resistive component and reactive component all in series are shown in Figure 7–11. With this example the coil has wire resistance as part of its opposition. Actually, the "wire resistance" is usually a very small part of the effective AC opposition in a coil. The coil has its own impedance and the Z triangle for the coil will be added to the circuit oppositions. Remember from Chapter 4 the methods for calculating the current, voltage, and power of a coil with both inductive reactance and resistance.

The same process will be used here with the addition of capacitors and resistors. We will use common values of components to help keep the mathematics easy while keeping track of circuit conditions. As we construct the impedance triangle for the circuit we need to pay attention to the vector directions and values for each component.

FieldNote!
....
To find the total impedance of a coil, a right triangle is constructed using X_L as the vertical component and using the resistance of the wire as the horizontal component. Refer to Figure 7-12. The hypotenuse of the triangle is the impedance Z of the coil. Refer to chapter 4 for more details.

FIGURE 7–11 Schematic of circuit with resistance, capacitance and coil with impedance.

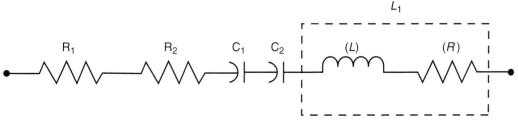

Example

Given:

$$E_T = 150 \text{ V}$$

$$R_1 = 50 \text{ } \Omega$$

$$R_2 = 50 \text{ } \Omega$$

$$X_{C1} = 50 \text{ } \Omega$$

$$X_{C2} = 50 \text{ } \Omega$$

Coil 1 has $X_L = 50 \text{ } \Omega$, R of coil $= 25 \text{ } \Omega$

FIGURE 7–12 Individual opposition vector diagrams for circuit in Figure 7–11.

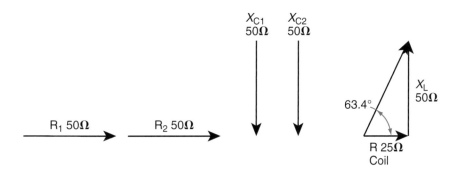

The individual opposition vectors appear as shown in Figure 7–12, and can be redrawn into a total impedance triangle as shown in Figure 7–13. As the vectors are added, the final Z triangle appears as in Figure 7–14. Note that each vector still represents an individual component with the impedance of the coil shown as 55.9 Ω of Z. The impedance of the circuit (Z_T) is 134.6 Ω.

FIGURE 7-13 Opposition vectors are all added in a series RLC circuit.

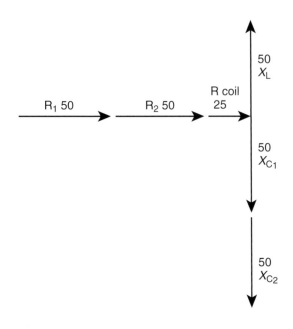

FIGURE 7-14 Total impedance triangle with calculations for individual components.

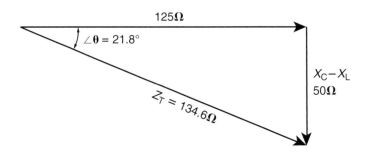

The triangle shows the resultant of all the components creates a more capacitive circuit and the current leads the line voltage by 21.8°. The line current is:

$$I_T = \frac{E_T}{Z_T}$$

$$I_T = \frac{150}{134.6} = 1.11 \text{ A}$$

To find the volt drop across each component, use the line current multiplied by the individual opposition.

• $E_{R1} = I_T \times R_1$

$$E_{R1} = 1.11 \times 50$$

$$E_{R1} = 55.5 \text{ V}$$

- $E_{R2} = I_T \times R_T$

$$E_{R2} = 1.11 \times 50$$

$$E_{R2} = 55.5 \text{ V}$$

- $E_{C1} = I_T \times X_{C1}$

$$E_{C1} = 1.11 \times 50$$

$$E_{C1} = 55.5 \text{ V}$$

- $E_{C2} = I_T \times X_{C2}$

$$E_{C2} = 1.11 \times 50$$

$$E_{C2} = 55.5 \text{ V}$$

- $E_{L1} = I_T \times Z_{\text{coil}}$

$$E_{L1} = 1.11 \times 55.9$$

$$E_{L1} = 62 \text{ V}$$

As we have discovered before, if you add all the voltages arithmetically your sum of 284 V is greater than the supply voltage, which is not possible. We must build a voltage triangle to prove these figures in the above example are accurate. See Figure 7–15 for the voltage vectors and the final voltage triangle. Even though each component can be measured with a voltmeter, each measurement including the 62 volts across the coil can be added in the vector diagram and the total resultant of all vectors is approximately 150 volts, which is the supply voltage. Our figures do not match exactly because of rounding off of answers.

FIGURE 7–15 Voltage triangle created by individual component voltage vectors.

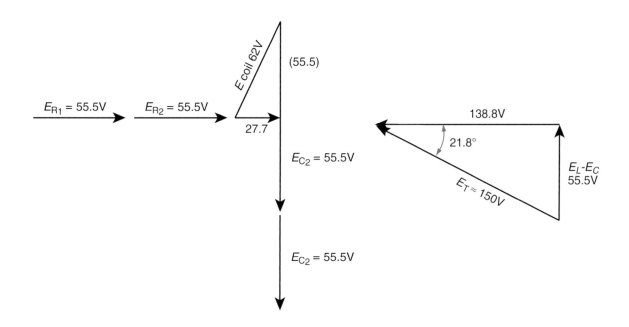

We can now construct the power triangle to verify calculated values. There are different methods to calculate the quantities desired. For instance, Watts can be calculated by using $I^2 \times R$ or $I \times E$, as applied in the first two calculations. Likewise, VARs or VA can be calculated by $I^2 \times$ opposition or $I \times E$ of the component. The methods are varied to show that different methods can be used.

- $W_{R1} = I^2 \times R_1$

$$W_{R1} = 1.11^2 \times 50$$

$$W_{R1} = 1.23 \times 50$$

$$W_{R1} \approx 61.5 \text{ W}$$

Note: Both W_{R1} and W_{R2} were found using two different methods. Either method could have been used for both.

- $W_{R2} = I_T \times E_{R2}$

$$W_{R2} = 1.11 \times 55.5$$

$$W_{R2} \approx 61.5 \text{ W}$$

- $VARs_{C1} = I_T \times E_{C1}$

$$VARs_{C1} = 1.11 \times 55.5$$

$$VARs_{C1} = 61.5 \text{ } VARs \text{ (capacitive)}$$

- $VARs_{C2} = I_T \times E_{C2}$

$$VARs_{C2} = 1.11 \times 55.5$$

$$VARs_{C2} = 61.5 \text{ } VARs \text{ (capacitive)}$$

- $VA_{L1} = I_T \times E_{coil}$

$$VA_{L1} = 1.11 \times 62$$

$$VA_{L1} = 68.8 \text{ VA}$$

- $W_{L2} = I^2 \times R_{coil}$

$$W_{L2} = 1.11^2 \times 25$$

$$W_{L2} = 1.23 \times 25$$

$$W_{L2} = 30.75 \text{ W}$$

- $VARs_{L1} = I^2 \times X_L$

$$VARs_{L1} = 1.11^2 \times 50$$

$$VARs_{L1} = 1.23 \times 50$$

$$VARs_{L1} = 61.5 \text{ VARs}$$

Note: VARS$_{L1}$ could also be solved using the same formula for VARs$_{C1}$.

As you can see from the preceding calculations the vectors for the power triangle are congruent with the other triangles. Adding all the watts together on the horizontal line yields 153.75 W, which is what a circuit watt meter would read. The coil has VA developed across the coil as well as VARs. The capacitors have only reactive power. See Figure 7–16 for the complete power triangle.

As you can see, the VA of the circuit should be 1.11 A times the 150 V supply, or 166 VA. The power triangle resultant vector is approximately 166 VA. The circuit phase angle is the same throughout all the triangles at 21.8° leading current. The power factor is the cosine of that angle or 92.8%. The power factor can be confirmed by using the $\frac{W}{VA}$ formula; in this case $\frac{154\ W}{166\ VA}$ is 92.8 %. Again, these figures are rounded off, to ease in the calculation.

FIGURE 7-16 Complete power triangle made up of individual calculated power components.

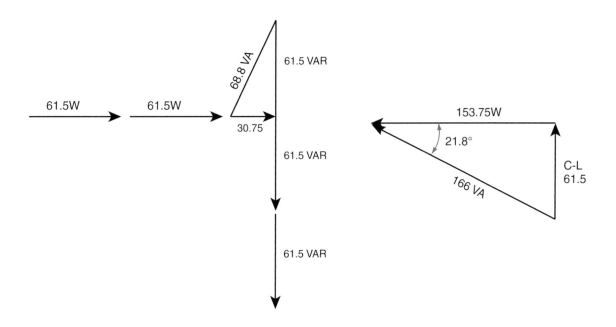

SUMMARY

The voltage drop across the resistor in a series RLC circuit is in phase with the current. The voltage drop across the inductor in a series RLC circuit will lead the current by 90°, and the voltage drop across the capacitor in a series RLC circuit will lag the current by 90°. The current is the same *amperage value* at all points in a series circuit.

Vector addition can be used in a series RLC circuit to find values of total voltage, impedance, and apparent power. In an RLC circuit, inductive and capacitive values are 180° out of phase with each other. The smaller value is subtracted from the larger value, resulting in a reduced larger value. Even though you can measure the component values independently, the total effects of the circuit must be added vectorially. All the total vector diagrams that create right triangles for oppositions, voltage, and power are congruent.

Series resonant circuits are used for various circuit needs, such as material detection circuits in automated manufacturing processes that are frequency sensitive. The circuits can be used in band pass filters, as discussed in later chapters, used to detect the presence of a desired frequency. In series resonance, the combination of the X_L and X_C will be equal at a specific frequency. At this exact point, the Z of the circuit is at a minimum and the circuit current is at a maximum.

Finally, we analyzed a circuit that introduced a more practical type circuit with a coil that had resistance rather than "theoretically" no wire resistance, which we used for some previous examples to build the process. This Z of the coil can be added into the series circuit and the results can be calculated.

REVIEW QUESTIONS

1. What are the voltage relationships across each of the components in a series RLC circuit?

2. What are the current relationships in a series RLC circuit?

3. When solving a series RLC circuit, describe how you would calculate the following:
   a. Total impedance
   b. Total current flow
   c. Resistive voltage
   d. Inductive voltage
   e. Capacitive voltage
   f. Power factor

4. Explain what happens to Z_T and I_T in RLC series circuit as the circuit frequency approaches resonance.

5. In a series RLC circuit operating at resonant frequency, what is the effect on circuit current?

6. Determine what the circuit power factor would be for a series RLC or LC circuit at resonant frequency.

7. How can you determine if RLC series circuit has a leading or lagging power factor by reviewing the impedance triangle from a mathematical standpoint?

8. What can you conclude about the frequency of a circuit when, no matter how you change the frequency, the wattmeter reads less than the original wattmeter reading?

9. In a series circuit containing a coil that has both a resistive and a reactive component, explain which component of the voltage triangle you will read with a voltmeter across the coil.

10. Write the formula used to calculate series resonant frequency.

PRACTICE PROBLEMS

Problems 1 to 6 refer to Figure 7–17.

1. Find the impedance.
2. Draw and label an impedance triangle for the series circuit as shown in Figure 7–17.
3. Find E_T.
4. Find E_R, E_C, and E_L.
5. What is the true power of the circuit?
6. What is the phase angle and power factor (PF)?

Problems 7 to 13 refer to Figure 7–18:

7. Draw and label an impedance triangle for the series circuit in Figure 7–18.
8. Find Z.
9. Draw and label a vector diagram of the circuit values for E_C, E_L, E_R, I_T, and E_T, and angle theta for the circuit in Figure 7–18.
10. Find I.
11. Find E_R, E_L, and E_C.
12. What power is used in this circuit? Prove it.
13. Find E_T and draw the vector diagram.

FIGURE 7–17   Circuit drawing for Problems 1–6.

FIGURE 7–18   Circuit drawing for Problems 7–13.

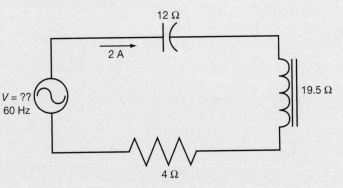

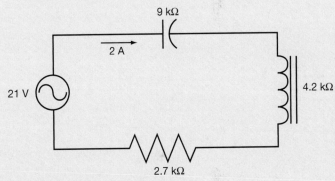

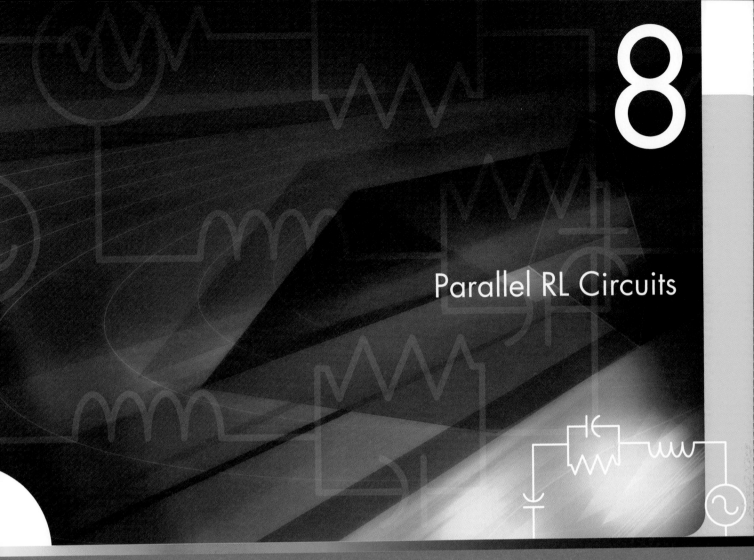

8

Parallel RL Circuits

OVERVIEW

We have connected coils in series with resistors. Now we move to AC parallel circuits. The process becomes a little more complicated than it was with parallel circuits in DC. The reason is that the currents and oppositions react differently to AC voltage. The current theoretically lags behind the voltage by 90° or somewhere between 0° and 90°. We will develop several different ways to solve the AC circuits with resistors and coils in parallel. After a quick review of parallel operation of resistors, coils will be added to the circuits. This chapter will add "pure coils" or coils with no wire resistance at first. Second explanations involve coils that have both X_L and R. We will proceed through current triangles, vectors, find impedance, and calculate power with a power triangle and also compute the power factor of the circuit.

OBJECTIVES

After completing this chapter, you should be able to:
- Calculate total resistance when you know the branch resistances
- Determine the total inductance of coils in parallel
- Calculate the total inductive reactance of coils in parallel
- Explain the process of determining total current and impedance of coils and resistors in parallel
- Calculate circuit quantities of coils in parallel with resistors to find the Z of the circuit
- Calculate branch and total current, impedance, voltage, and power for any R_L parallel circuit
- Determine power factor of an R_L parallel circuit

REVIEW OF PARALLEL CIRCUIT LAWS

A good review of the rules for parallel resistors and the methods of solving circuit calculations is appropriate. Remember that as resistors are added in parallel we are actually adding more paths for current to flow and therefore are actually decreasing the total opposition of the circuit. If we were to vector this type of circuit, the process would be to add all the current vectors, at the same angle, together to find the total circuit current. From that current and the line voltage that was applied, we could compute the circuit opposition.

The formula is the standard Ohm's law $R_T = \dfrac{E_T}{I_T}$ to find the circuit totals. We will use this same concept of adding current paths and adding the currents together to yield the total current. The formula for R_T with known resistors is:

$$\frac{1}{R_T} = \frac{1}{R_1} + \frac{1}{R_2} + \frac{1}{R_3} + \cdots \frac{1}{R_n}$$

or

$$R_T = \cfrac{1}{\dfrac{1}{R_1} + \dfrac{1}{R_2} + \dfrac{1}{R_3} + \cdots \dfrac{1}{R_n}}$$

There are many variations of the formula to calculate R_T if all the resistors are the same, using product over sum, assuming a voltage, etc. We will not review all of these rules now.

INDUCTORS CONNECTED IN PARALLEL

When inductors are connected in parallel, the total inductance can be found using a formula that is similar to the one used for finding total resistance in a parallel circuit. The reciprocal of the total inductance is equal to the sum of the reciprocals of all the inductors (Figure 8–1). Calculations for Figure 8–1 are as follows:

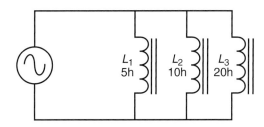

FIGURE 8–1   Three coils in a parallel AC circuit can be combined for a total L.

$$\frac{1}{L_T} = \frac{1}{L_1} + \frac{1}{L_2} + \frac{1}{L_3} + \cdots \frac{1}{L_n}$$

or

$$L_T = \cfrac{1}{\dfrac{1}{L_1} + \dfrac{1}{L_2} + \dfrac{1}{L_3} + \cdots \dfrac{1}{L_n}}$$

$$L_T = \cfrac{1}{\dfrac{1}{L_1} + \dfrac{1}{L_2} + \dfrac{1}{L_3}} = \cfrac{1}{\dfrac{1}{5} + \dfrac{1}{10} + \dfrac{1}{20}}$$

$$L_T = \frac{1}{.2 + .1 + .05} = \frac{1}{.35} = 2.85 \text{ H}$$

The same rules apply that we see for parallel resistors, that the total inductance of the circuit is smaller than the smallest branch inductance. The total inductance formula for two inductors in parallel is formatted the same way as the one for two resistors in parallel:

$$L_T = \frac{L_1 \times L_2}{L_1 + L_2}$$

Example

Two coils are connected in a circuit in parallel. They have an inductance of 10 H and 5 H. What is the total inductance of the two in parallel? Use the product-over-sum method.

Solution:

$$L_T = \frac{5\ H \times 10\ H}{5\ H + 10\ H}$$

$$L_T = 3.33\ H$$

If all inductors are the same value, use the concepts that the total inductance will be the inductance of one coil divided by the number of coils as below.

$$L_T = \frac{L_C}{N}$$

where:
L_T is the circuit total inductance
L_C is the common inductance
N is the number of like inductors

INDUCTIVE REACTANCE IN PARALLEL

Because inductance and inductive reactance are directly proportional, the same rules that apply for inductors in parallel are also used to determine inductive reactance in parallel. In other words, if the total inductance goes up then the reactance increases. If the total inductance decreases, the total reactance decreases.

Adding additional inductors in parallel with a circuit that already contains inductors will decrease the total inductive reactance of the circuit. This follows the same principle that adding additional resistor paths in parallel reduces the total resistance. Just as with resistors, the total opposition to current flow in the circuit containing parallel inductors is always less than the opposition caused by the lowest-value coil or inductor because each time another element is added in parallel, the current has another path. Therefore, the same rules apply for inductive reactance as for resistors:

$$X_{L_T} = \frac{1}{\dfrac{1}{X_{L_1}} + \dfrac{1}{X_{L_2}} + \dfrac{1}{X_{L_3}} + \cdots \dfrac{1}{X_{L_n}}}$$

or for two reactive components:

$$X_{L_T} = \frac{X_{L_1} \times X_{L_2}}{X_{L_1} + X_{L_2}}$$

Example

Find the total inductive reactance of the circuit in Figure 8–2 with two coils that are each .133 H and are operating at 60 Hz.

- $X_{L_1} = 2\pi fL$

  $X_{L_1} = 2\pi 60(.133)$

  $X_{L_1} = 50 \ \Omega$

- $X_{L_2} = 2\pi fL$

  $X_{L_2} = 2\pi 60(.133)$

  $X_{L_2} \approx 50 \ \Omega$

- $XL_T = \dfrac{XL_1 \times XL_2}{XL_1 + XL_2}$

  $XL_T = \dfrac{50 \times 50}{50 + 50}$

  $XL_T = 25 \ \Omega$

Or use the total inductance as:

- $L_T = \dfrac{L_C}{N}$

  $L_T = \dfrac{.133}{2}$

  $L_T = .0665 \ \text{H}$

  $XL_T = 2\pi fL_T$

  $XL_T = 2\pi(60)(.0665)$

  $XL_T = 25 \ \Omega$

FIGURE 8–2

Two identical coils in parallel can be reduced to an equivalent several ways.

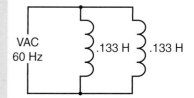

VAC 60 Hz .133 H .133 H

FieldNote!

Or use the formula for identical opositions as in

$X_{LT} = \dfrac{X_{L\ common}}{n}$

Where:
$X_{L\ common}$ = the value of the common coil
n = number of common coils

$X_{LT} = \dfrac{50}{2}$

$X_{LT} = 25 \ \Omega$

PURE INDUCTORS AND RESISTORS IN PARALLEL

For circuits that contain resistors and inductors in parallel, the concept is the same as resistors in parallel but the mechanics of calculating the totals are a little more involved. We will first begin with a coil and a resistor in parallel. The coil as shown in Figure 8–3 is a "perfect coil" having no resistance, only inductive reactance. We will begin the process with this scenario to build understanding. Unlike two resistors in parallel, to find the total opposition we will calculate the total current that will flow. As you know the current in a perfect coil will lag the voltage of the coil by 90 electrical degrees. As you also know about parallel circuits, the voltage is the same across all the branches. Voltage is our constant and will be the reference drawn on the 0° axis. All currents that are "in phase" with the line voltage are also referenced to the 0° axis. Therefore, the current that flows through the resistive branch is in phase with the applied voltage.

FIGURE 8–3

Resistor and inductor in parallel.

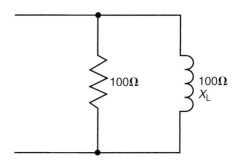

100Ω   100Ω X_L

The current flow through the 100-Ω resistor of 1 A $\left(I = \dfrac{E}{R}\right)$ is represented by I_R in Figure 8–4. The current through the 100 Ω of inductive reactance (X_L) is 1 A $\left(I = \dfrac{E}{X_L}\right)$ and is drawn at 90° behind the voltage reference or drawn downward and labeled I_L. Be sure to label these vectors, as they can get confusing depending on whether you're drawing I_L or X_L in series or in parallel.

Now that we have two vectors at 90° angles to each other, we can add them vectorially to obtain the resultant, or hypotenuse of 1.414 A at a 45° angle to the horizontal reference.

The total current of this circuit is 1.414 A, and it lags the line voltage by 45°. To find the impedance of this circuit use $Z_T = \dfrac{E_T}{I_T}$. In our circuit the total Z is $\dfrac{100\ \text{V}}{1.414\ \text{A}} = 70.7\ \Omega$ of impedance. We now have solved for the values for the current triangle for the circuit in Figure 8–4.

FIGURE 8–4   Current vectors representing the magnitude and direction of the current in a RL parallel circuit.

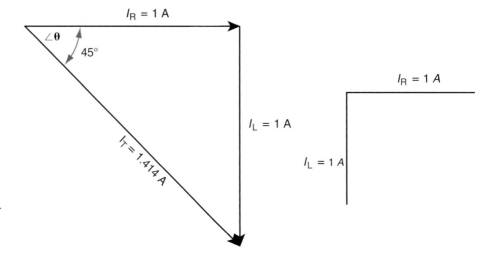

As you can see there is more than one way to provide calculations. One method is more mathematical, using just algebra, while the other uses the visual aid of drawing vectors. Using the vectors seems less confusing as we add more components to a typical circuit and also create circuits with R, X_L, and X_C, then finally all three components in combinations of series and parallel. Although you will need to use the system that is most comfortable for you, the vector addition methods you have already used will be presented first.

Another method to accomplish the same result is to use the product-over-sum method used in parallel resistors for two components at a time. This process works well for two components but is a little more cumbersome for multiple branches. The end result is to know the same quantities but not the phase angle. The formula is as follows:

- $Z_T = \dfrac{X_L \times R}{\sqrt{(X_L^2 + R^2)}}$

  $Z_T = \dfrac{100 \times 100}{\sqrt{(10,000 + 10,000)}}$

  $Z_T = \dfrac{10,000}{141.4}$

  $Z_T = 70.7\ \Omega$

- $I_T = \dfrac{E_T}{Z_T}$

  $I_T = \dfrac{100\ \text{V}}{70.7\ \Omega}$

  $I_T = 1.414\ \text{A}$

FIGURE 8–5 Schematic of resistor in parallel with a coil containing R and X_L.

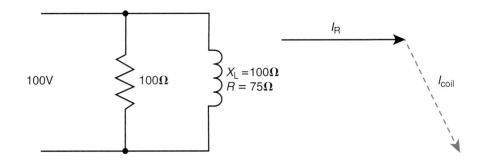

100V   100Ω $X_L = 100\Omega$ $R = 75\Omega$

I_R

I_{coil}

COILS WITH IMPEDANCE IN PARALLEL WITH RESISTORS

Now the process becomes more involved as we add a "real coil" that contains both R and X_L to a parallel resistor as shown in Figure 8–5. With the addition of the coil resistance of the wire added to the inductive reactance of the coil we actually create a series circuit in parallel with the 100 Ω resistance. Let us use the same reactance value from our first example, namely 100 Ω of X_L but add 75 Ω of R to the coil circuit. This wire resistance effective resistance value is an exaggerated value to help follow the process. We now have to rely on your skill in series circuit calculations to determine the current flow in the coil branch of Figure 8–5. Refer to Figure 8–6 to see the impedance created by the coil. The Z of the coil is 125 Ω. The angle created by the total impedance vector is 53.1° lagging, or more inductive.

FIGURE 8–6 Coil branch has Z determined by opposition vectors resistance and reactance.

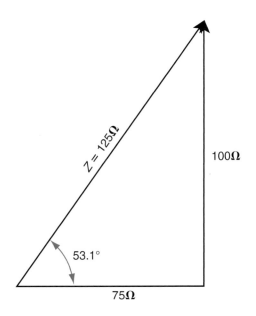

$Z = 125\Omega$ 100Ω

53.1°

75Ω

FieldNote!

When using trigonometry to solve AC calculations and add vectors, be aware of formulas for sine, cosine, and tangent. In Figure 8–6 you will need to know the angle created by the hypotenuse (**Z**) and the **R** of 75 Ωs. Calculate this angle with the tangent function, which is the ratio of opposite side to adjacent side or $\frac{O}{A}$. $T = \frac{O}{A}$ will produce a decimal dividend that can be entered into a calculator to find the arc (angle) that has a tangent of $\frac{100}{75} = 1.3333$; the angle with a tangent of 1.3333 is 53.1°. You can use trigonometry to find the hypotenuse too, by using the angle of 53.1° and one of the sides. Use either the $\sin = \frac{O}{H}$ or the $\cos = \frac{A}{H}$ to solve for the Hypotenuse (H). The hypotenuse can also be solved by using the two sides and Pythagorean's theorem.

FieldNote!

When adding vectors at various angles, it is essential to construct vectors so that we can find the horizontal and vertical components. In Figure 8–7, the current vector of .8 A is at an angle to the horizontal of 53.1°. To add this to the I_R of 1 A, use the function of $\cos = \dfrac{A}{H}$ from trigonometry to find the adjacent side. The cosine of 53.1° is .6; .6 $= \dfrac{A}{.8}$, so the Adjacent side is .48 A at 0° as shown. This triangle's vertical side is found with the sine function. The sin of 53.1° $= .799$, so .799 $= \dfrac{O}{.8}$, therefore the opposite, or vertical component is .639. To get the resultant of the combination of the two original currents, add all horizontal components to get 1.48 A, and create a right triangle with the .639 A of the vertical component. The hypotenuse can be calculated by using trigonometry or the Pythagorean theorem.

Remember that this impedance has an effect on the current through that component, causing the current to lag the voltage by the same 53.1°. Now if we calculate the current through the coil it is $I_{\text{coil}} = \dfrac{E_{\text{coil}}}{Z_{\text{coil}}}$ or $I = \dfrac{100\ \text{V}}{125\ \Omega}$ or $I = .8$ A

Unlike the first example, the current through the coil is no longer theoretically 90° lagging but is really 53.1° behind the voltage. Referring back to the vector addition of currents in parallel, we must now add the current vectors together (Figure 8–7). As we add the vectors, one at 0° (the current through the resistor) and one at a 53.1° lag (the current through the coil), the resultant will be the circuit current and the resultant angle will be the circuit voltage and current displacement. The resultant is 1.61 A at 23.4° lag. The line current (1.61 A) lags the line voltage (100 V) by 23.4°. The percent power factor for the circuit is the cosine of this angle theta (θ) × 100 or .918 × 100 = 91.8% power factor. We will confirm this later in this chapter.

FIGURE 8–7   Vector addition of currents at various angles.

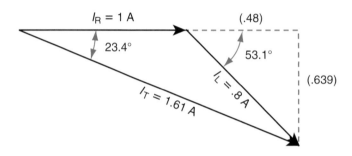

CURRENT, IMPEDANCE, VOLTAGE, AND POWER

As you calculate the quantities involved you need to track the angles, the vectors for the individual branches, and the power that is developed at each component. Be sure to verify that these individual values still add to the total and that the total makes sense with the applied source. For this verification example process, we will again choose to ignore the resistance of the coil, meaning that it is a high Q of 10, or quality factor as referenced in Chapter 4—the reactance is at least 10 times the resistance. See Figure 8–8 for this example

FIGURE 8–8   Two resistors and two "perfect" high Q coils in parallel.

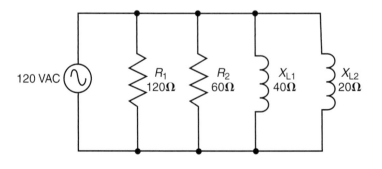

of two coils and two resistors. As you notice the circuit values are designed to aid in understanding and not to be cumbersome in mathematical calculations. Assuming we use 120 V AC 60 Hz and the values of 120 Ω and 60 Ω of resistance and .106 H 40 Ω X_L and .053 H 20 Ω X_L, respectively we can easily calculate the total current flow from the source. Use the $X_L = 2\pi fL$ and 60 Hz to calculate the X_L of each coil. See Figure 8–9 for the total current triangle based on the current flow through each branch. Even though the branches are 1 A, 2 A, 3 A, and 6 A, respectively, the total is *not* the arithmetic sum of 12 A. Instead, the resultant hypotenuse of the triangle is 9.48 A. Through trigonometry we can calculate the circuit phase angle as 71.6 °.

FIGURE 8–9 Current triangle resulting from Figure 8–8.

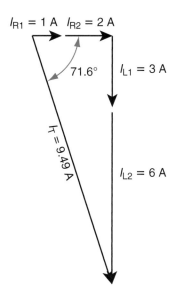

$I_{R1} = 1$ A $I_{R2} = 2$ A

71.6° $I_{L1} = 3$ A

$I_T = 9.49$ A

$I_{L2} = 6$ A

8.1 Impedance of the Circuit

The impedance of the circuit could be calculated by using the product of the oppositions divided by the square root of the sum of the squares of the opposition, two components at a time. A more efficient manner is to use the vector sum of the branch current and divide by the source voltage as follows:

$$Z_T = \frac{E_T}{I_T}$$

$$Z_T = \frac{120 \text{ V AC}}{9.48 \text{ A}}$$

$$Z_T = 12.66 \ \Omega$$

For reference this means that we have a total of 12.66 Ω of impedance at a 71.6° lagging angle to the purely resistive horizontal reference.

FieldNote!

Power can only be dissipated or consumed by a resistor; therefore, there is nothing to add up unless it involves a resistor. The total power can be added mathematically by the power each resistor consumes; otherwise it is VARs or VA, which is very different. Theoretically, there is no power consumed by either an inductor or a capacitor. The only other power we have in an AC circuit is *Apparent Power* or $VA = E_{applied} \times I_T$.

8.2 Voltage of the Circuit

As is the rule for a parallel circuit in DC circuits as well as AC circuits, the voltage value is consistent for all branches of the circuit and it remains at 120 V AC. The current and the phase angle through each component are dictated by the opposition encountered in each branch. See Figure 8–8 for voltage verification.

8.3 Power in this Circuit

Power in the parallel circuit is dependant on the individual branches. Each branch acts as an independent load. Unlike the loads that are all in phase with each other as in a parallel DC circuit, we are not allowed to add the loads arithmetically unless they all have the same phase angle. The power measurements must be added based on their phase angle relationship with the voltage and current. In this instance, the voltage across each opposition is the same and the current is based on the individual opposition. Therefore, to find the power we need to take the branch current times the branch voltage. Power for the resistive components will be in the form of watts. Power for the reactive components will be in VARs. Then each power component is vectorially added to yield the circuit totals. See the following calculations.

- $W_{R1} = E \times I_1$

  $W_{R1} = 120 \text{ V} \times 1 \text{ A}$

  $W_{R1} = 120 \text{ W}$

- $W_{R2} = E \times I_2$

  $W_{R2} = 120 \text{ V} \times 2 \text{ A}$

  $W_{R2} = 240 \text{ W}$

- $VARs_{L_1} = E \times I_{L_1}$

  $VARs_{L_1} = 120 \text{ V} \times 3 \text{A}$

  $VARs_{L_1} = 360 \ VARs$

- $VARs_{L_2} = E \times I_{L_2}$

  $VARs_{L_2} = 120 \text{ V} \times 6 \text{ A}$

  $VARs_{L_2} = 720 \ VARs$

With these vectors we can complete the power triangle as seen in Figure 8–10. The resultant VA, or hypotenuse of the triangle, is 1138 VA. To confirm these calculations we can take the line current times the line voltage to get circuit volt-amps of apparent power:

$$120 \text{ V} \times 9.48 \text{ A} \approx 1138 \text{ VA}$$

FIGURE 8–10 Power triangle resulting from Figure 8–8.

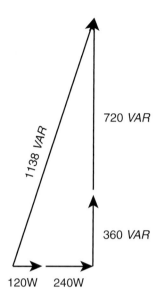

The last step is to confirm the phase angle and the power factor. The power factor is still the $\left(\dfrac{\text{watts}}{\text{volt-amps}}\right) \times 100$ equals percent power factor (*PF*). We can also use the cosine of the phase angle between circuit current and circuit voltage $\times$ 100 = % *PF*. See the following calculations.

$$\% \ PF = \left(\frac{\text{W}}{\text{VA}}\right) \times 100$$

$$\% \ PF = \left(\frac{360 \ \text{W}}{1138 \ \text{VA}}\right) \times 100$$

$$\% \ PF = .316 \times 100$$

$$\% \ PF = 31.6 \ \%$$

or

$$\% \ PF = \text{cosine } 71.6° \times 100$$

$$\% \ PF = .3156 \times 100$$

$$\% \ PF = 31.6\%$$

Parallel Impedance Formula

Still another way to solve the last example is to add the resistors in parallel as you would with a DC circuit to get an equivalent circuit. 120 Ω and 60 Ω in parallel create an equivalent value of 40 Ω.

FieldNote!

The type of parallel circuit that is most prevalent is the parallel circuit with multiple resistive values and some reactive values. A typical home may have a 20 A 120 V kitchen circuit with multiple loads. In parallel could be a toaster, an electric can opener, a waffle iron, and an electric knife. Each of these loads can create its own branch of a parallel circuit with varying degrees of phase angle. The circuit can be calculated at the circuit breaker as the voltage source to determine how this AC circuit would react.

FIGURE 8-11   Equivalent circuit of two coils and two resistors—Parallel Impedance Formula.

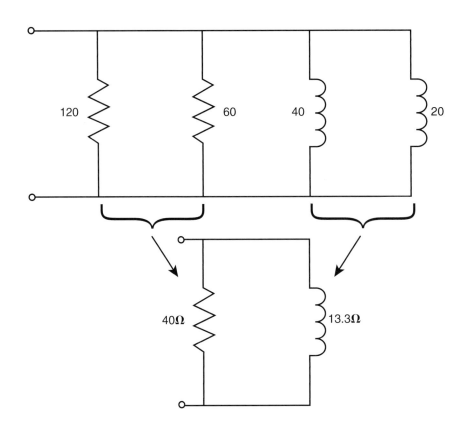

Then the coils of 40 Ω and 20 Ω could be added together to get an X_L of 13.3 Ω. See Figure 8–11 for equivalent circuit. The final Z is the same as we calculated in earlier examples. We don't have the convenience of all the branch currents to quickly find power and it is harder to double check our results, but the results are the same when we use high Q coils where we disregard the resistance in the coil circuits.

$$Z = \frac{(R \times X_L)}{\sqrt{(X_L{}^2 + R^2)}} \quad \text{(Parallel Impedance Formula)}$$

$$Z = \frac{40 \times 13.3}{\sqrt{(13.3^2 + 40^2)}}$$

$$Z = \frac{533}{\sqrt{(176.9 + 1600)}}$$

$$Z = \frac{533}{42.2}$$

$$Z = 12.6$$

MULTIPLE COILS WITH WIRE RESISTANCE AND RESISTORS IN PARALLEL

FIGURE 8-12 Two different value resistor and two different value coils in parallel.

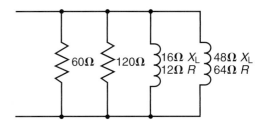

In this last example we will use two coils and two resistors each with a different value. This might be similar to the kitchen circuit mentioned in the previous sidebar. Figure 8–12 provides an example. Again, the values may be exaggerated to show method and process rather than actual appliance values. A 60 Ω (R1) and a 120 Ω (R2) resistor are in parallel with coil 1 (L1) which has 16 Ω X_L and 12 Ω R and coil 2 (L2) which has 48 Ω X_L and 64 Ω R. The circuit voltage is 120 V. You can solve this problem using the following equations:

$$I = \frac{E}{R} \text{ (for restrictive branches)}$$

$$I = \frac{E}{Z} \text{ (for inductive branches)}$$

- Find $I_{R1} = \dfrac{E_T}{R_1} = \dfrac{120 \text{ V}}{60 \text{ Ω}} = 2 \text{ A at } 0°$

- $I_{R2} = \dfrac{E_T}{R_2} = \dfrac{120 \text{ V}}{120 \text{ Ω}} = 1 \text{ A at } 0°$

  Find Z of coil by $Z = \sqrt{R^2 + X_L^2}$, then $I = \dfrac{E}{Z}$. Remember to find the phase angle of the impedance of each branch using the trigonometric functions with R and X_L as the sides of the right triangle.

- $I_{L1} = \dfrac{E_T}{L_1} = \dfrac{120 \text{ V}}{20 \text{ Ω}} = 6 \text{ A at } 53.1°$

- $I_{L2} = \dfrac{E_T}{L_2} = \dfrac{120 \text{ V}}{80 \text{ Ω}} = 1.5 \text{ A at } 36.9°$

- Add these current vectors (Figure 8–13).

FIGURE 8-13   Vector addition for currents for circuit 8–12.

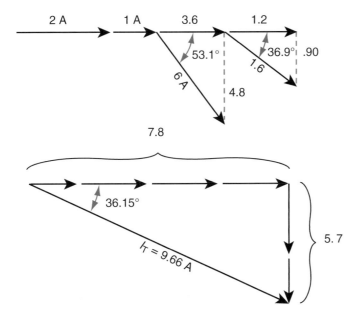

The currents for each branch were found in equations for Figure 8–12. The two coil currents were at an angle to the horizontal, matching the impedance triangle for each coil that you would have calculated in Figure 8–12. The third branch had a coil with a phase angle of 53.1° and the last branch had a coil with a phase angle of 36.9°. By using the branch currents at the appropriate phase angles, we can construct a vector addition using a head-to-tail method. By using trigonometry, the horizontal side of the 6 A vector at 53.1° is 3.6, and the vertical component of the same 6 A vector is 4.8. The fourth branch has a current of 1.5 A at an angle of 36.9°. The horizontal component is 1.2 and the vertical component is .9. We can now construct the large current triangle with all the horizontal and vertical components as shown. Using the Pythagorean theorem, you will find the hypotenuse as 9.66 A. The angle between the horizontal voltage reference and the hypotenuse is calculated by trigonometric functions of Sin, Cos, or Tan and the known sides. The circuit phase angle is 36.15°.

The vector sum is 9.66 A at 36.15° to the horizontal and is lagging because it is an inductive circuit.

- The voltage is a constant 120 V.
- The power triangle will be congruent to the current triangle as in Figure 8–14.

FIGURE 8–14 Power triangle for Circuit 8-12.

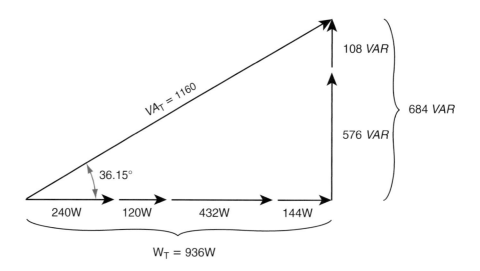

True power is measured in watts and is represented or shown along the horizontal axis because it has a 0° angle. The *VARs* are represented or shown along the vertical axis caused by and are a result of the reactive components. The VA or apparent power is a product of the line current times the line voltage and also forms the hypotenuse of the power triangle. The final figures are rounded off to ease the calculation process but maintain the theory.

$W = E \times I$ (branch voltage × branch current)

- W in $R_1 = 120 \text{ V} \times 2 \text{ A}$
  W in $R_1 = 240 \text{ W}$

- W in $R_2 = 120$ V $\times$ 1 A

  W in $R_2 = 120$ W

$W = I \times E$ (horizontal component of branch current triangle $\times$ branch voltage)

- W in $L_1 = 3.6$ A resistive current $\times$ 120 V

  W in $L_1 = 432$ W

- W in $L_2 = 1.2$ A resistive current $\times$ 120 V

  W in $L_2 = 144$ W

$VARs = I \times E$ (vertical component of branch current triangle $\times$ branch voltage)

- $VARs$ in $L_1 = 4.8$ A reactive $\times$ 120 V

  $VARs$ in $L_1 = 576$ VARs

- $VARs$ in $L_2 = .9$ A reactive $\times$ 120 V

  $VARs$ in $L_2 = 108$ VARs

$W_T = W_1 + W_2 + W_3 + W_4$

- Total watts $= 240 + 120 + 432 + 144 = 936$

$VARs$ total $= VAR_1 + VAR_2$

- Total $VARs = 576 + 108 = 684$ $VARs$

VA total = line voltage $\times$ line current

- Total $VA = 9.66 \times 120$ V $= 1160$ VA

- Power factor $= \left(\dfrac{936 \text{ W}}{1160 \text{ VA}} \right)$

  % $PF = 80.7\%$

  Cosine of angle $36.15° \times 100 = 80.7\%$

SUMMARY

In this chapter we have created circuits with coils in parallel with resistors. Because the coils and resistors have different phase angles in an AC circuit, we have found that they cannot be inserted into an equation in the same way as we did with DC circuits.

One method of calculating Z was to vectorially add the current vectors and find a total current, and then use total E and total I to get total Z. There are other methods that can be applied under special circumstances. The process of calculating all the circuit variables also included the calculation of AC power which could be illustrated by the power triangle. This led to the power factor and the angle of lag, which was the same angle represented by a current triangle.

REVIEW QUESTIONS

1. Explain why total resistance decreases as we add more resistors in parallel.
2. What is the relationship between total inductance and branch circuit inductance in parallel circuits?
3. Write a formula for finding the X_{LT} of three different coils in a parallel AC circuit.
4. Explain a method of adding a branch resistor current to a separate branch inductor current in order to find the total circuit current.
5. Explain why the phase angle of a current triangle parallel current vector would be the same as a power triangle for the same circuit.

6. How would you calculate the voltage across a parallel circuit if you only knew one coil X_L and the current through that coil?
7. What is meant by adding a "pure" inductance in parallel?
8. What is the power factor of a circuit that has 30° lagging inductive current behind the resistive current?
9. If a circuit has 10 A line current and 200 V line voltage, but only consumes 150 W with a watt meter, what is the reactive power ($VARs$)?
10. Find the percent power factor (% PF) for the circuit in Question 9.

PRACTICE PROBLEMS

All practice problems refer to Figure 8–15.

1. What is the current through the resistive branch?
2. What is the current through the inductive branch?
3. Find I_T.
4. Construct a current vector diagram.
5. What is the impedance of this circuit?
6. Show how to find the impedance in Problem 5 using the parallel impedance formula.
7. Find the total circuit power factor (PF) of the RL circuit.
8. A decrease in frequency in a parallel circuit will cause the PF to _____ .

FIGURE 8-15   Circuit for Practice Problems 1–8.

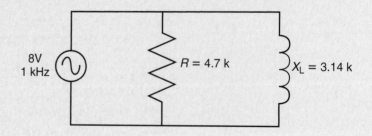

8V
1 kHz

$R = 4.7\ k$

$X_L = 3.14\ k$

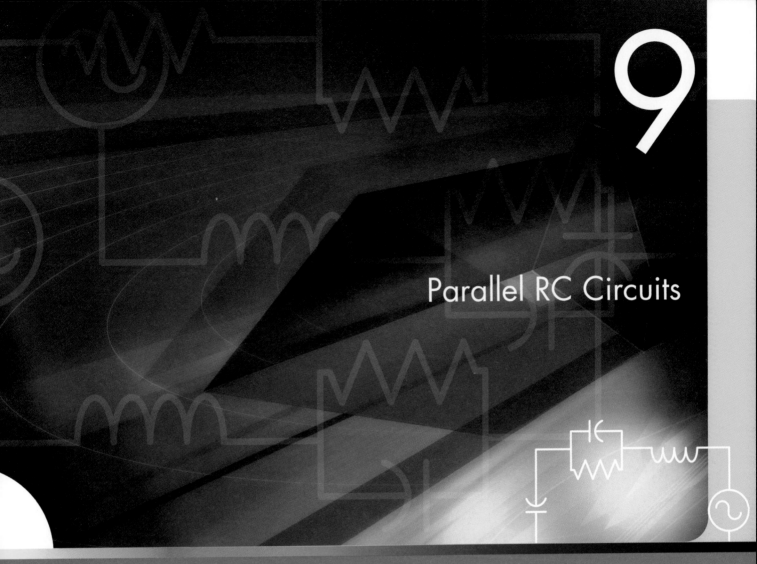

9

Parallel RC Circuits

OVERVIEW

The calculations for solving parallel capacitor circuits are similar to those of parallel resistance and inductance circuits. The primary differences are that capacitors in series are calculated like resistors and inductors in parallel and that capacitors in parallel are calculated like resistors and inductors in series.

In Chapter 6, the series RC circuit was discussed. Remember that in series circuits, current is the same through all the components, and is the reference point. The voltages are in relation to the reference current.

In a parallel circuit, voltage is the same across all the components. In parallel circuits that contain both resistance and reactance, the branch currents are out of phase, and the voltage is the reference. Again, a parallel circuit that contains resistance and capacitance is called an RC circuit. In this chapter we will analyze how the capacitive currents combine with the resistive currents and affect the circuit totals. We will determine the impedance of the circuit based on the total current flow rather than based on vector additions of the oppositions provided by resistors and capacitors. Finally we will determine what the circuit trends are as we change variables in the RC parallel circuit.

OBJECTIVES

After completing this chapter, you should be able to:
- Determine the effects of capacitors in parallel
- Calculate the total capacitance of a circuit with multiple capacitors in parallel
- Calculate the capacitive reactance of parallel capacitors
- Find all required circuit quantities such as current, impedance, voltage, power
- Perform circuit calculations for all RC parallel circuits
- Describe the effects of changing parameters in a parallel RC circuit
- Determine mathematically unknown quantities in these circuits

CAPACITORS IN PARALLEL

As first discussed in Chapter 6, we found the effects on the circuit when capacitors were connected in series. When the capacitors are connected in parallel, the areas of the individual plates are added together, increasing the capacitance. Connecting capacitors in parallel has the same effect as increasing the plate area of one capacitor. Look at Figure 9–1 to equate the plate size to parallel capacitors. The formula used to determine the capacitance when the dielectric constant (K), the area of the plates (A) (in square inches), and the distance between the plates (d) (in inches) also applies here. You can see that increasing the surface area of the plates increases the capacitance value (C). Capacitors in parallel are calculated similar to inductors and resistors in series.

FIGURE 9–1   The effect of parallel capacitors equate to larger plate area in the capacitance formula.

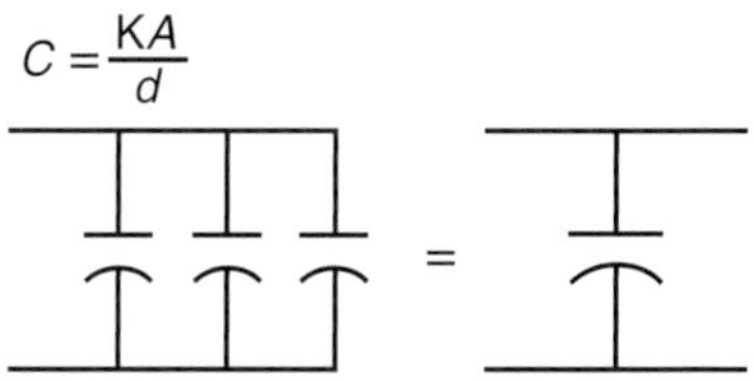

$$C = \frac{KA}{d}$$

(a) Gets Larger

Previously, you learned that when *inductors* are connected in *series,* the total induction of the circuit equals the sum of the inductances of each of the inductors:

$$L_T = L_1 + L_2 + L_3 + \ldots L_n \quad \text{(series)}$$

The total resistance in a series circuit is found by adding the values of the resistors in the circuit:

$$R_T = R_1 + R_2 + R_3 + \ldots R_n \quad \text{(series)}$$

Connecting capacitors in parallel has the same effect as increasing the plate area of one capacitor. Capacitors in parallel are calculated similar to inductors and resistors in series.

The total capacitance of a parallel capacitance circuit is given by:

$$C_T = C_1 + C_2 + C_3 + \ldots C_n \quad \text{(parallel)}$$

FIGURE 9–2   Three separate capacitors in a parallel AC circuit.

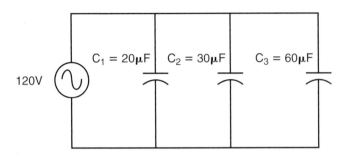

Example

Calculate the total capacitance of the circuit shown in Figure 9–2.

Solution:

$$C_T = C_1 + C_2 + C_3$$

$$C_T = 20 \ \mu F + 30 \ \mu F + 60 \ \mu F$$

$$C_T = 110 \ \mu F$$

CAPACITIVE REACTANCE IN PARALLEL

Capacitive reactance is an opposition to the flow of AC current as discussed in Chapter 5. As such, we can treat the reactance of the capacitor as another ohmic value as we would ohms of resistance or ohms of inductive reactance. We can add the ohmic values of components that have the same reaction to the phase relationship in the AC circuit. As we know, the ohmic value is actually due to the voltage developed by the capacitor charge and creates a value that is 90° out of phase with the resistance. Therefore, when we add capacitive reactance in parallel we just use the same rules as resistors or inductive reactance in parallel. By adding more parallel paths we actually increase the number of current paths, and reduce the total opposition of the circuit. We would use the following formulas for finding total X_C when we know the individual capacitive reactance of the circuit.

$$\frac{1}{X_{CT}} = \frac{1}{X_{C1}} + \frac{1}{X_{C2}} + \frac{1}{X_{C3}} \cdots \frac{1}{X_{CN}}$$

or

$$X_{CT} = \frac{1}{\dfrac{1}{X_{C1}} + \dfrac{1}{X_{C2}} + \dfrac{1}{X_{C3}} \cdots \dfrac{1}{X_{CN}}}$$

or, product over sum for two branches at a time:

$$\frac{X_{C1} \times X_{C2}}{X_{C1} + X_{C2}}$$

or, if all branches are alike and *N* represents the number of like branches then:

$$\frac{X_C}{N}$$

or, if you know the C_T then $X_{CT} = \dfrac{1}{2 \ \pi f C_T}$ can be used

Example

Calculate the total capacitive reactance of Figure 9–3 by using the individual capacitance values or the individual reactances at 60 Hz.

Solution:
Reactances in parallel behave the same as resistances in parallel. Therefore, the total reactance is

- $C_T = C_1 + C_2 = C_3$

  $C_T = 20 \ \mu F + 30 \ \mu F + 60 \ \mu F$

  $C_T = 110 \ \mu F$

- $X_{CT} = \dfrac{1}{2\pi f C_T}$ Note: C_T is in microfarads (μF)

  $X_{CT} = \dfrac{1}{(377)(110 \ \mu F)}$

  $X_{CT} = 24.1 \ \Omega$

or

- $X_{C1} = \dfrac{1}{2\pi f C_T} = \dfrac{1}{(377)(20 \ \mu F)}$

  $X_{C1} = 132.6 \ \Omega$

- $X_{C2} = \dfrac{1}{2\pi f C_T} = \dfrac{1}{(377)(30 \ \mu F)}$

  $X_{C2} = 88.4 \ \Omega$

- $X_{C3} = \dfrac{1}{2\pi f C_T} = \dfrac{1}{(377)(60 \ \mu F)}$

  $X_{C3} = 44.2 \ \Omega$

- $X_{CT} = \dfrac{1}{\dfrac{1}{X_{C1}} + \dfrac{1}{X_{C2}} + \dfrac{1}{X_{C3}} \cdots}$

  $X_{CT} = \dfrac{1}{.0.00754 + .0113 + .0226}$

  $X_{CT} = \dfrac{1}{.0414}$

  $X_{CT} = 24.1 \ \Omega$

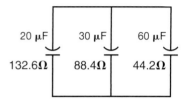

FIGURE 9–3

Three capacitors with capacitive reactance values.

RC PARALLEL CIRCUITS

When resistance and capacitance are connected in parallel, the voltage across all the devices will be in phase and will have the same value. The current flow through the capacitor will be 90° out of phase, or leading the capacitive voltage. The phase angle shift between the total current and total voltage is determined by the ratio of the amount of resistance to the amount of capacitive reactance. Since the amount of capacitive reactance and amount of resistance will independently determine the amount of current flowing through each branch, whether it is the resistive or capacitive branch, the total current flow can be determined only after those values are determined.

The capacitive branch of the circuit is directly affected by the capacitive reactance which is dependant on the capacitance value and the frequency of the source.

CIRCUIT VALUES

In Figure 9–4, the resistor (25 Ω) is connected in parallel with a capacitor that has a capacitive reactance of 20 Ω. The circuit power supply is 220 V at a frequency of 60 Hz. Remember that the resistive and capacitance branches will have to be calculated separately, and then combined for total circuit values.

FIGURE 9-4   Parallel circuit of Resistance and Capacitance.

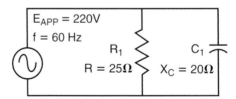

9.1 Circuit Currents

Refer to Figure 9–4 for the following calculations. The current flow through the resistor can be calculated using the following formula:

$$I_R = \frac{E}{R}$$

$$I_R = \frac{220 \text{ V}}{25 \text{ Ω}}$$

$$I_R = 8.8 \text{ A}$$

FIGURE 9–5   Current vector triangle diagram for circuit in Figure 9–4.

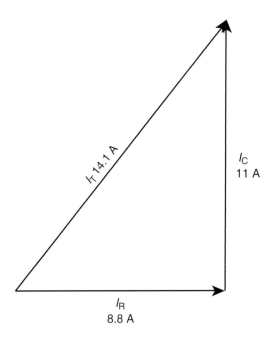

The current flow through the capacitor is calculated using the following formula:

$$I_C = \frac{E}{X_C}$$

$$I_C = \frac{220 \text{ V}}{20 \text{ } \Omega}$$

$$I_C = 11 \text{ A}$$

In this simple circuit we can determine the circuit current by adding the currents. Because the currents are not in phase with each other they are not occurring at the exact same point in time, and so we need to add them using the right triangle of the Pythagorean theorem or by vector addition. See Figure 9–5 for the triangle developed by adding the resistive current with the capacitive current that is 90° out of phase. As you know, the addition of 8.8 A and 11 A does not equal 19.8 A but instead yields the resultant current of approximately 14.1 A. This resultant current is the current delivered by the circuit from the source of supply. You can also view the current triangle vector to find the angle of lead—the angle between the horizontal axis and the hypotenuse. We use trigonometry to calculate the arc which has a tangent of ratio between the opposite and the adjacent sides to the angle. In this case the formula for the current phase angle is as follows:

$$\text{Tangent } \theta = \frac{\text{opposite side magnitude}}{\text{adjacent side magnitude}} = \frac{I_C}{I_R}$$

$$\text{Tan } \theta = \frac{11 \text{ A}}{8.8 \text{ A}}$$

Find the arc which has a tangent of 1.25

Arc tan 1.25 = θ of 51.3°

FieldNote!

The actual current read by an ammeter placed in the branch circuits of the circuit will actually read a resistive current of 8.8 A, a capacitive current of 11 A, and a total current of 14.1 A. To find actual circuit component values we can use Ohm's law to determine the opposition based on the current and the voltage. If we do not know the capacitor value, we can calculate the value by measuring the current I_C and using the formula $X_C = \dfrac{E}{I_C}$ to find the X_C. Then working backwards from that value we use the formula of $C = \dfrac{1}{2\pi f X_C}$ to find the capacitor value in farads.

We now can tell that the circuit current leads the circuit voltage by 51.3°. We can verify this fact as we work through the rest of the calculations.

Another way to get the total current with these two currents that are 90° out of phase with each other is to use the Pythagorean formula as follows:

$$I_T = \sqrt{I_R^2 + I_C^2}$$
$$I_T = \sqrt{8.8^2 + 11^2}$$
$$I_T = \sqrt{77.44 + 121}$$
$$I_T = 14.1 \text{ A}$$

9.2 Circuit Impedance

The total circuit impedance can be found using any of the total values and substituting Z for opposition in the Ohm's law formula. Remember that circuit impedance is not found by vector addition of the individual oppositions. In parallel circuits, the current paths are the deciding criteria for the total opposition of the circuit. We need to find the total current and then determine what impedance caused that current flow. The total impedance can be calculated using the formula:

$$Z_T = \frac{E_T}{I_T}$$

$$Z_T = \frac{220 \text{ V}}{14.1 \text{ A}}$$

$$Z_T = 15.6 \text{ }\Omega$$

VOLTAGE

The circuit voltage in a parallel circuit stays the same throughout each branch of the circuit as it does in any parallel circuit. We need to remember that the current and the voltage in a resistive branch is "in phase" or reaching peak and zero points of the respective sine wave at the same point in time. However, the current and the voltage for the capacitive branch are 90° out of phase. When we need to verify the voltage across the resistor, it is still $I \times R = E$. The voltage across the capacitor is $I \times X_C = E_C$. In the previous example in Figure 9–4, we can calculate the voltage across the capacitor as 11 A $\times$ 20 Ω = 220 V. When we consider the circuit values in total, we know that the circuit current is somewhere between 0 and 90° leading the circuit voltage. In the first example the current is leading the voltage by 53.1° because it is a capacitive effect circuit.

POWER IN RC PARALLEL AC CIRCUITS

Remember that true power is consumed in the resistive parts of a circuit where current flow is in phase with the voltage. The amount of true power in the circuit can be determined by using any of the values associated with the purely resistive part of the circuit. For Figure 9–4, true power can be found using the following formula:

$$P = E \times I_R$$

$$P = 220 \times 8.8 \text{ A}$$

$$P = 1936 \text{ W}$$

9.3 Apparent Power

The apparent power can be computed by multiplying the circuit voltage times the total current flow. This is the hypotenuse of the power triangle as shown in Figure 9–6:

$$VA = E \times I_T$$

$$VA = 220 \text{ V} \times 14.1 \text{ A}$$

$$VA = 3102 \text{ VA}$$

9.4 Reactive Power

In this example, energy is stored by the capacitor in the form of volts of pressure and coulombs of electric charge in the electrostatic field. The power that is delivered to the circuit is measured in volts-amps-reactive or VARs−C. We need to determine the amount of capacitive current and the amount of voltage across the capacitor circuit. These two values multiplied together will yield the VARs of the RC circuit. In the example circuit we calculated the current through the capacitor at 11 A and the capacitor voltage is 220 V. See Figure 9–6 for the total power triangle. The VARs formula is as follows:

$$VARs = E_{\text{line}} \times I_{\text{line}}$$

$$VARs = 220 \text{ V} \times 11 \text{ A}$$

$$VARs = 2420 \text{ VARs}$$

Verify these values by using the power triangle as shown in Figure 9–6.

FieldNote!

In parallel RC circuits, the voltage measured across each parallel component appears to be the same value. However, the voltages are out of phase with each other, which is not seen with a voltmeter. The vectors would look like this:

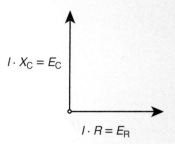

$I \cdot X_C = E_C$

$I \cdot R = E_R$

These vectors are not added together.

TechTip!

As is the case in all AC circuits, the line current times the line voltage will yield apparent power. Only in completely resistive circuits is the apparent power measured in VA equal to the true power measured in watts. As we will see, we can always use the formula of $E \times I \times$ power factor to yield true power in an AC circuit. Sometimes the power factor is 1.00 (100%). The power factor can never be more than 100% and is only 100% for totally resistive circuits.

FIGURE 9–6 Power triangle for circuit in Figure 9–4.

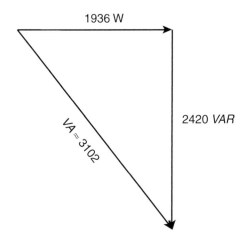

1936 W

2420 VAR

VA ≈ 3102

POWER FACTOR

As is the case in all AC circuits the power factor (*PF*) is a factor applied to the apparent power of the circuit. As we multiply the decimal equivalent of the %*PF* of the circuit, to the apparent power of the circuit, the product will yield the true power. The power factor is the ratio of true power to apparent power. One formula for determining the power factor is the following:

$$PF = \frac{\textit{True power in watts}}{\textit{Apparent power in VA}}$$

$$\% \ PF = \left(\frac{W}{VA}\right) \times 100$$

$$\% \ PF = \left(\frac{1936 \ \text{W}}{3102 \ \text{VA}}\right) \times 100$$

$$\% \ PF = 62.4\%$$

ANGLE THETA (θ)

Remember that the cosine of angle theta (θ) is equal to the power factor. The angle θ is the angle that the current is out of phase with the voltage. In the power triangle it is the angle determined by the ratio of Watts to Apparent Power (*VA*). The formula is as follows:

$$\cos \angle\theta = \frac{\text{adj}}{\text{hyp}} = \frac{W}{VA} = \frac{1936}{3100} = .624$$

$$\% \ PF = \cos\theta \times 100 = 62.4\%$$

$$\cos^{-1} .624 = \angle\theta$$

$$\text{angle } \theta = 51.3°$$

Remember this angle is the same angle found in the current triangle of 51.3° leading current. This sample problem has a leading power factor of 62.4%.

MULTIPLE RESISTORS AND CAPACITORS IN PARALLEL

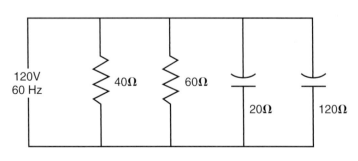

FIGURE 9–7 Schematic diagram of a RC circuit with two capacitors and two resistors.

120V
60 Hz

40Ω   60Ω

20Ω   120Ω

Multiple resistors and capacitors in parallel is not as difficult as true coils (High Q) and resistors in parallel. Because capacitors have such little resistance compared to their capacitive reactance, we do not need to include the *R* of the capacitor as we calculate the actual branch currents. Therefore, the addition of more resistors or capacitors is a relatively easy extension of the rules that were just presented. We will need to vector current of the resistors at 0° and currents of the capacitors at 90° leading. We will find all the circuit parameters for the circuit as shown in Figure 9–7.

Example

To keep the mathematics to a minimum, we will use simple values. The steps are the same as we have previously used for parallel AC circuits: (1) find the branch currents; (2) add the appropriate values to create a circuit current triangle; (3) determine the circuit impedance if needed; (4) calculate the phase angle of the currents; (5) find the power components; (6) construct a circuit power triangle to determine each power component. In our simple circuit we have two resistors rated at R1 @ 40 Ω and R2 @ 60 Ω and two capacitors with X_{C1} @ 20 Ω and X_{C2} @ 120Ω. The supply voltage is 120 V AC @ 60 Hz. Use the following formulas:

- $I_{R1} = \dfrac{E}{R_1}$

  $I_{R1} = \dfrac{120}{40 \ \Omega}$

  $I_{R1} = 3 \ A \ @ \ 0°$

- $I_{R2} = \dfrac{E}{R_2}$

  $I_{R2} = \dfrac{120 \ V}{60 \ \Omega}$

  $I_{R2} = 2 \ A \ @ \ 0°$

- $I_{C1} = \dfrac{E}{X_{C1}}$

  $I_{C1} = \dfrac{120 \ V}{20 \ \Omega}$

  $I_{C1} = 6 \ A \ @ \ 90°$

- $I_{C2} = \dfrac{E}{X_{C2}}$

  $I_{C2} = \dfrac{120 \ V}{120 \ \Omega}$

  $I_{C2} = 1 \ A \ @ \ 90°$

Construct a circuit triangle with the currents as shown in Figure 9–8. Find the resultant vector by using the Pythagorean theorem. The resultant current of 8.6 A is the current from the source, or I_T. The line current leads the line voltage by approximately 54.5°.

The impedance of the circuit is:

$$Z_T = \frac{E_T}{I_T}$$

$$Z_T = \frac{120 \ V}{8.6 \ A}$$

$$Z_T \approx 14 \ \Omega$$

FIGURE 9–8

Current triangle for the branch currents found in Figure 9–7.

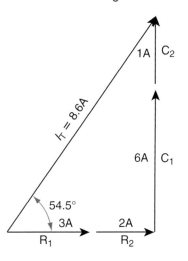

The power components are as follows as depicted in Figure 9–9.

- $W_{R1} = E \times I_{R1}$

  $W_{R1} = 120 \text{ V} \times 3 \text{ A}$

  $W_{R1} = 360 \text{ } W$

- $W_{R2} = E \times I_{R2}$

  $W_{R2} = 120 \text{ V} \times 2 \text{ A}$

  $W_{R2} = 240 \text{ W}$

- $VARs_{C1} = E \times I_{C1}$
- $VARs_{C1} = 120 \text{ V} \times 6 \text{ A}$
- $VARs_{C1} = 720 \text{ VARs}$

  $VARs_{C2} = E \times I_{C2}$

  $VARs_{C2} = 120 \times 1 \text{ A}$

  $VARs_{C2} = 120 \text{ VARs}$

FIGURE 9–9   Power triangle of vectors for circuit in Figure 9–7.

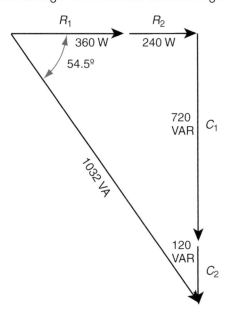

The final power triangle as shown in Figure 9–9 will verify that the figures all coincide with the actual measured values if the circuit were constructed. The total *VA* according to the Pythagorean theorem is 1032.3 VA. This should be the same as the line current times the lines voltage as follows:

$$VA = \text{Line E} \times \text{Line I}$$

$$VA = 120 \text{ V} \times 8.6 \text{ A}$$

Apparent power = 1032 *VA*, the same as with the Pythagorean theorem

The angle θ is arc tan of $\dfrac{VARs}{W}$

$$\text{Arc tan } \theta = \frac{840 \text{ VARs}}{600 \text{ W}}$$

$$\text{Arc tan } \theta = 1.4$$

$$\text{Arc} = 54.5°$$

This angle matches the phase angle of the current triangle vector.

The %*PF* is the cosine of this angle θ $\times$ 100. The cosine of 54.5° is .58 $\times$ 100 = 58% *PF*. To verify this figure with another method calculate

PF as $\left(\dfrac{W}{VA}\right) \times 100 = \%PF$. Again, using circuit values of $\left(\dfrac{600 \text{ W}}{1032 \text{ VA}}\right) \times 100$

the *PF* is approximately 58%. The figures do not exactly match because we rounded many of the values.

EFFECTS ON CHANGING CIRCUIT VARIABLES

As the circuit components or variables change so do the circuit values. As shown in Table 9–1, changing the frequency will change the X_C and a host of other dependant variables. This table will indicate whether the values change or stay the same and what the trends are with the subsequent measurable quantities. Try to determine the effects on the circuit then verify the trend by checking the table.

Refer to Table 9–1 for effects of frequency, capacitance, or resistance on RC parallel circuits when changing one variable at a time.

TechTip!

Oftentimes we do not need to know the exact values of a circuit to troubleshoot the circuit. We can determine what is generally happening to the circuit by the readings we take and then change one parameter of the circuit to see if it responds in a predicted manner. If we suspect that a capacitor is open, we can add another capacitor to the parallel circuit to see if the circuit reacts as we suspect. Table 9–1 allows us to see the effects of changing one variable at a time.

TABLE 9–1   Trends in an RC Parallel Circuit When Changing One Variable at a Time

Variable	X_C	I_C	I_R	I_L	Z_L	VA	P true (w)	VARs	Phase Angle	PF
Freq Inc	Dec	Inc	Same	Inc	Dec	Inc	Same	Inc	Inc	Dec
Freq Dec	Inc	Dec	Same	Dec	Inc	Dec	Same	Dec	Dec	Inc
Cap Inc	Dec	Inc	Same	Inc	Dec	Inc	Same	Inc	Inc	Dec
Cap Dec	Inc	Dec	Same	Dec	Inc	Dec	Same	Dec	Dec	Inc
Resis Inc	Same	Same	Dec	Dec	Inc	Dec	Dec	Same	Inc	Dec
Resis Dec	Same	Same	Inc	Inc	Dec	Inc	Inc	Same	Dec	Inc

Inc = increase; Dec = decrease

SUMMARY

This chapter introduced you to the effects of capacitors in parallel and how to calculate the circuit quantities based on the capacitance or the capacitive reactance. A resistor and a capacitor in a parallel circuit created the effect of adding possible current paths for the circuit current although the resistive current and the capacitive current were not in phase with each other. This led us to current vectors and current triangles. The phase angle for the circuit could be determined by the angle theta (θ) found in the current triangle and later confirmed in the power triangle. The impedance of the circuit was determined based on the total current. The voltage stayed the same in all branches of the circuit as it does in all parallel circuits and is our circuit reference. Finally, the power triangle was constructed to determine all the power measurements as well as confirming the power factor.

Table 9–1 was used to determine how the circuit would change as we changed variables in the RC parallel circuit.

REVIEW QUESTIONS

1. What is the effect of connecting capacitors in parallel?
   a. What does the size of the cross-sectional area of the plates affect?
   b. How does this affect the total capacitance?
2. Briefly describe how to calculate the total capacitance of capacitors in parallel and show formula.
3. Briefly describe how to calculate the total capacitive reactance of capacitors in parallel and show formula.
4. Describe how to find the following in an RC parallel circuit:
   a. Phase angle
   b. Power factor
   c. Resistive voltage drop
   d. Reactance voltage drop

5. Discuss impedance in a parallel RC circuit.
   a. How can it be calculated?
   b. How will it compare to resistance and capacitative reactance?
6. Draw a vector diagram for an RC parallel circuit and label all components.
7. By adding more capacitors in parallel the total X_C for the circuit will _____.
8. Increasing the frequency of the RC circuit would cause the *VARs* to _____.
9. The hypotenuse of the current triangle is a measurable current known as the _____ current.
10. Adding more resistors in parallel causes the circuit impedance to _____.

PRACTICE PROBLEMS

1. What value of capacitor must be added in parallel with a .23 μF capacitor to get a total of 5000 Ω of X_C total in a 60 Hz, 1000 V circuit?

2. What is the capacitive reactance of two 50 μF capacitors is parallel with 100 Hz applied?

3. A 5 V circuit at 100 kHz has a 1000% PF capacitor in parallel with a 1000 Ω resistor. Find I_R.

4. Find X_C.

5. Find I_C.

6. Find I_T.

7. Find Z.

8. Find the phase angle.

9. Find PF.

10. What is the apparent power delivered to this circuit?

11. What is the true power in this circuit?

12. In the same circuit, will an increase in frequency to 200,000 Hz cause this circuit to be more or less capacitive? Prove your answer.

13. Find the total capacitance of three capacitors connected in series with values of 3 μF, 5 μF, and 10 μF.

14. What is the total capacitance of two capacitors connected in parallel having values of 310 μF and 50 μF?

15. What is the X_{CT} of a circuit having two capacitors connected in parallel with values of 300 μF and 600 μF operating at 500 Hz?

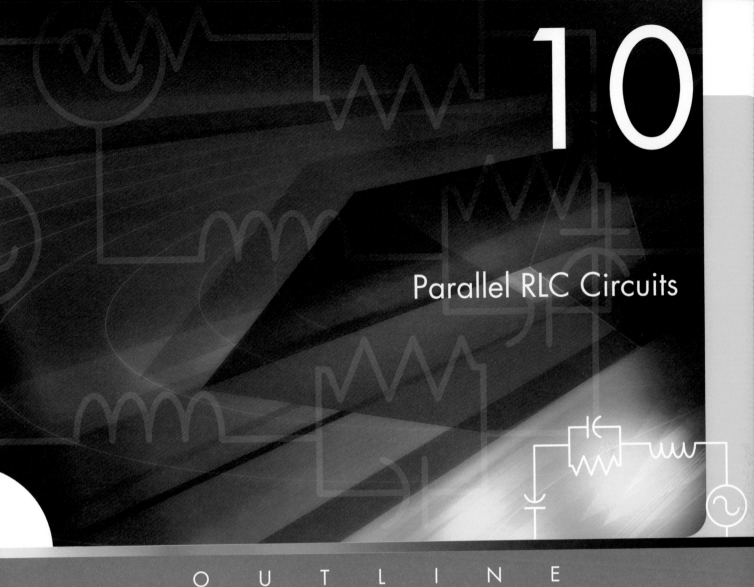

10

Parallel RLC Circuits

OVERVIEW

Most of the circuits you will work with in your professional career will contain some combination of resistance, inductance, and capacitance. Most AC loads tend to be resistive with inductance because of the heavy presence of inductive loads like motors and fluorescent light fixtures. Further, most of the circuits you will work with will be a combination of series and parallel circuits. The load circuits themselves are usually primarily either R or RL in series and are connected in parallel between the supply wires.

This chapter continues the work you did in the previous chapters. Remember that in a parallel circuit, the voltage is the same across all branches, and the current divides among the branches. We will investigate how changing variables in the circuit will affect the total circuit. These variations resemble the way the circuit changes each time you add or subtract an AC load to an existing circuit. We will also explore resonance and how the resonant point of the circuit affects the power and the line current and explain what is meant by a "tank" circuit in parallel AC circuits with capacitance (C) and inductance (L).

Also in this chapter we will use an "assumed voltage" to calculate the impedance of an AC parallel circuit. This is a way to determine the impedance of an RLC circuit without knowing what the voltage will actually be. We can use this method of calculation to temporarily calculate the total current of a circuit without knowing the exact voltage present. This concept can be useful to determine values in a complex circuit where exact values are hard to decipher.

OBJECTIVES

After completing this chapter, you should be able to:
- Determine the overall effects of parallel AC components
- Calculate for unknown quantities in a RLC circuit
- Calculate total current and total impedance in RLC circuits with alternate methods
- Find the power for RLC circuits with given variables
- Determine power factor and circuit phase angles
- Determine circuit impedance using an assumed voltage
- Explain the effect of a RLC circuit at resonance

PARALLEL RLC CIRCUITS

When an AC circuit has the elements of resistance, inductance, and capacitance connected in parallel, the voltage drop across each branch is the same as it is in any parallel circuit. The magnitude and phase angle of the currents flowing through each branch are out of phase with each other as determined by the resistance or reactance of the branch.

Referring to Figure 10–1, the phase relationship of each branch current in parallel is represented by the relationship of current to voltage in the individual branch. As you recall, the current in a pure resistive component would be in phase with the component voltage. In the pure capacitive circuit the component current leads the component voltage by 90°, and the current in the pure inductive component would be 90° behind the component voltage. In the total parallel circuit with all three branches, the line current and line voltage may be anywhere from 90° leading current, to 0° (or in phase) current, to 90° lagging current behind the applied circuit voltage.

FIGURE 10–1 Current relationships in different branches of an RLC circuit.

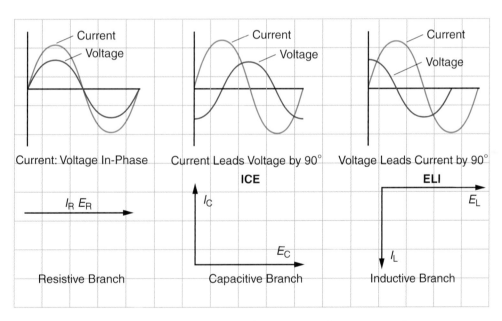

In a parallel RLC circuit, if the inductance is large, so is the X_L, causing a comparatively small amount of current to flow through the inductive branch. If the capacitive reactance is small, more of the circuit current flows through the capacitive branch. The circuit current would appear to have capacitive effects, and the line current would lead the line voltage. As we will determine, the effects of changing values in a parallel RLC circuit can depend on the frequency. The circuit may change from a leading to a lagging power factor circuit depending on how close it originally is to resonance.

CALCULATING PARALLEL RLC CIRCUIT VALUES

Use Figure 10–2 as a reference. The following problems are solved using the circuit values shown on the drawing. We can analyze this circuit as we did the other parallel circuits. The most systematic way to solve these is to determine the individual branch circuit characteristics and then add them into the total circuit equation. Remember that the currents in the branches will have amperage that you can read with a standard ammeter, but they also have a phase angle in relation to the branch voltage. Also recall that the voltage drop is the same across any component in a parallel circuit and is the main reference to our calculations.

FIGURE 10–2   Schematic of RLC parallel circuit for subsequent calculations.

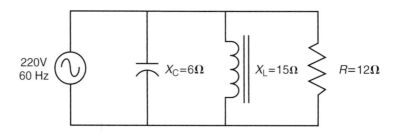

First: Calculate each current as follows:

10.1 Resistive Current

The easiest value to calculate is the current flow through the resistor. This can be calculated using the Ohm's law formula:

$$I_R = \frac{E}{R}$$

$$I_R = \frac{220 \text{ V}}{12 \text{ } \Omega}$$

$$I_R = 18.3 \text{ A @ } 0°$$

10.2 Inductive Current

The amount of current flow through the inductor is calculated using the adapted Ohm's law formula. Remember that in this "perfect" inductor there is no resistive component and the current lags the voltage by 90 electrical degrees.

$$I_L = \frac{E}{X_L}$$

$$I_L = \frac{220 \text{ V}}{15 \text{ } \Omega}$$

$$I_L = 14.67 \text{ A @ } 90° \text{ lag}$$

10.3 Inductance

We can calculate the amount of inductance in the circuit to verify the comparative values for henries vs. farads as we equate the effects of this circuit later. The formula to use is the inductive reactance formula as follows:

$$X_L = 2\pi f L$$

$$L = \frac{X_L}{2\pi f}$$

$$2\pi f = 377 \text{ at } 60 \text{ Hz, therefore}$$

$$L = \frac{15}{(377)}$$

$$L = 0.397 \text{ H}$$

10.4 Capacitive Current

The remaining branch of the circuit is capacitive. Therefore, the current through this component can be measured with a standard ammeter but you realize that the current actually leads the branch voltage by 90 electrical degrees. The current flow through the capacitor can be found using the adapted Ohm's law formula:

$$I_C = \frac{E}{X_C}$$

$$I_C = \frac{220 \text{ V}}{6 \text{ } \Omega}$$

$$I_C = 36.67 \text{ A @ } 90° \text{ lead}$$

10.5 Capacitance

For circuit comparison, we can find the actual capacitor reactance of the capacitive branch for comparison to the inductive branch values. The value of circuit capacitance opposition can be calculated using the capacitive reactance formula when the capacitance (C) is in farads.

$$X_C = \frac{1}{2\pi f C}$$

$$C = \frac{1}{2\pi f X_C}$$

$$C = \frac{1}{377 \times 6} = 0.000442 \text{ Farads}$$

$$C = 442 \text{ } \mu F$$

10.6 Impedance

Impedance for this circuit can be calculated in two different ways. The simplest way is to keep track of the currents and then to find the total current and then calculate the opposition (impedance) that caused that circuit current.

A second way is to use the mathematical formula that adds the opposition of each branch. Remember from previous chapters that the impedance of the parallel circuit is the reciprocal of the sum of the reciprocals of the branches. Vector addition must be used because the values are out of phase: The formula below assumes that the reactance of two branches is combined into an equivalent reactance by using the product-over-sum method. Then the reactance and resistance can be used in a modified product-over-sum formula where the denominator is the square root of the sum of the squares. This modification is needed because the resistance and reactance are 90° out of phase.

First combine two branch reactance values:

$$X_T = \frac{X_L \times X_C}{X_L + X_C}$$

Then, circuit impedance is calculated by the formula:

$$Z_T = \frac{X_T \times R}{\sqrt{X_T{}^2 + R^2}}$$

Important: When solving for X_T, the X_L is a positive number, and X_C is a negative number. It is vitally important that this concept be followed as shown here for the circuit in Figure 10–2.

Example

$$X_T = \frac{15 \times -6}{15 + -6}$$

$$X_T = \frac{-90}{9}$$

$$X_T = -10$$

The minus value shows that the net result is a capacitive reactance. Because the next part of the calculation uses the Pythagorean theorem, the negative sign will not be used:

$$Z_T = \frac{X_T \times R}{\sqrt{X_T{}^2 + R^2}}$$

$$Z = \frac{10 \times 12}{\sqrt{10^2 + 12^2}}$$

$$Z = \frac{120}{\sqrt{100 + 144}}$$

$$Z = \frac{120}{\sqrt{244}}$$

$$Z = \frac{120}{15.6}$$

$$Z = 7.68 \ \Omega$$

This final circuit impedance figure matches the figure found when vectoring currents and creating a current triangle vector diagram then using Ohm's law: $Z = \dfrac{E}{I}$.

TOTAL CIRCUIT CALCULATIONS

10.7 Total Current

The value for total current flow in the circuit can be calculated using vector addition of the current flow through each branch of the circuit. Refer to Figures 10–3a and b. As shown in Figure 10–3a, the inductive current is 180° out of phase with the capacitive current. These two currents tend to offset each other (subtract) the same way opposing reactances did in the series circuit. You will subtract the smaller from the larger. The total circuit current is the hypotenuse of the resultant right triangle.

TechTip!

The reactive branch that has the largest current flow determines if the line current will lead or lag the applied voltage, and will determine the phase angle. The phase angle can be found by:

$$\text{Angle Theta} = \tan\theta = \frac{R}{X}$$

FIGURE 10–3   Vector diagrams of branch currents for Circuit 10–2.

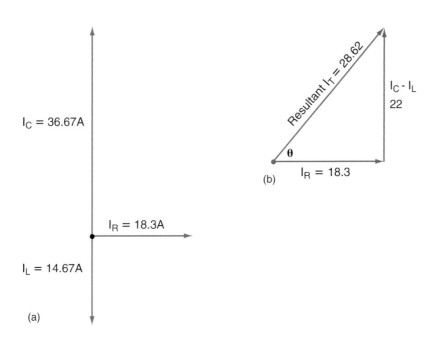

FieldNote!

A clamp-on style ammeter is the typical way an electrician measures AC current in a circuit. By measuring the magnetic field around a conductor, the meter converts this measurement to amperes of current flow. We can measure the current at many different points such as at the source of the current, at the resistive branch, or at the reactive branches. The meter only measures the amount of magnetic field and cannot determine if the current is in phase with the voltage. If you simply add all of the AC currents together, as we did in DC parallel circuits, the sum total may exceed the supply current reading. Keep this in mind as you measure currents in parallel circuits where you may not know for sure what type of load is connected for various branches.

The following steps show how the total current in Figure 10–3B could be calculated using the Pythagorean theorem:

$$I_T{}^2 = I_R{}^2 + (I_C - I_L)^2$$

$$I_T = \sqrt{I_R{}^2 + I_C - I_L{}^2}$$

$$I_T{}^2 = (18.32)^2 + (36.67 - 14.67)^2$$

$$I_T{}^2 = 334.89 + (22)^2$$

$$I_T{}^2 = 344.89 + 484$$

Note: The total current (I_T) formula to the right is often stated as:

$$I_T = \sqrt{I_R{}^2 + (I_C - I_L)^2}$$

$$I_T{}^2 = 818.89$$

$$I_T = \sqrt{818.89}$$

$$I_T = 28.62 \text{ A}$$

The same answer can be calculated using the same total current triangle and using trigonometric functions.

From this information we know the total current of the circuit as measured in Figure 10–4 with an ammeter as shown will be 28.6 A. As you know the supply current is not in phase with the supply voltage. We can calculate the phase angle of lead by using the current triangle and finding angle theta (θ). By using trigonometry and the final circuit current triangle from Figure 10–3, we can determine the angle between the horizontal axis, representing the voltage reference, and the total circuit current of I_T, which is the hypotenuse of the right triangle. The trigonometric function that is easiest to use in this case is the cosine function: $\text{Cos } \theta = \dfrac{\text{adjacent side}}{\text{hypotenuse}}$. $\text{Cos } \theta = \dfrac{18.3}{28.62}$ or $\text{Cos } \theta = .6394$. The angle is found by using the arc-cos or the $\cos^{-1}$ function to find the angle that has a cosine of .6394, which is 50.2°. For this circuit the angle is 50.2°.

FIGURE 10–4   Ammeters inserted in circuit the same as Figure 10–2.

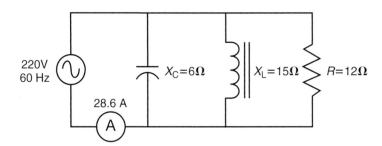

10.8 Alternate Way to Calculate Total Current

A second way to calculate total current is to use the impedance value as shown below:

$$I_T = \frac{E}{Z}$$

$$I_T = \frac{220}{7.68\ \Omega}$$

$$I_T = 28.65 \text{ A}$$

Note the slight difference in the current values due to rounding.

POWER CALCULATIONS

There are several ways to calculate the power of the original circuit with inductance, capacitance, and resistance all in parallel. One method is to construct a power triangle that will be congruent to the current triangle using the current and voltage of each branch to build the power components.

10.9 True Power

Remember that true power can be found only in the resistive leg of the circuit where current and voltage are in phase. The true power or watts can be calculated using the formula: $P = E \times I_R$. Another way is to use the $P = I^2 \times R$, where we must use only the branch current through the resistive branch. A third method of calculating the true power is to construct a power triangle. We will verify the watts (horizontal vector) of that triangle as the true power.

$$P = E \times I_R$$

$$P = 220 \times 18.33 \text{ A}$$

$$P = 4032.6 \text{ W}$$

Note: you can also use

$$P_{\text{True}} = E_{\text{APP}} \times I_T \times \cos\theta$$

$$PF = \cos\theta = \frac{Z}{R}$$

or

$$P = I^2 \times R$$

$$P = 18.33^2 \times 12 \text{ } \Omega$$

$$P = 4031.8 \text{ W}$$

A slight difference occurs when rounding repeating decimals.

10.10 Reactive (VAR) Power

The reactive (Volt-Ampere-Reactance) power is made up of both VARs of inductive power and VARs of capacitive power. We will use the inductive branch current and the circuit voltage (the applied voltage) to calculate the volts-amps-reactive (VARs) See Figure 10–5 as a representation of the VARs inductive vector from the formula

$$\text{VARs}_L = I_L \times E_L$$

$$\text{VARs}_L = 14.7 \times 220 \text{ V}$$

$$\text{VARs}_L = 3234 \text{ } VARs \text{ inductive}$$

FIGURE 10–5 Volt-amps-reactive vectored as capacitive or inductive VARs.

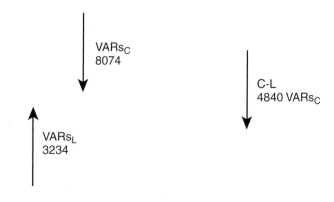

The next branch that creates reactive VARs is in the capacitive branch with VARs C is as follows:

$$VARs_C = I_C \times E_C$$

$$VARs_C = 36.7 \times 220 \text{ V}$$

$$VARs_C = 8074 \text{ VARs capacitive}$$

As you note in Figure 10–5 the inductive and capacitive VARs subtract because they are in opposite directions and the net (difference) in VARs is: $VARs_C$ @ 8074 − $VARs_L$ @ 3234 = 4840 $VARs_C$, or more capacitive.

10.11 Apparent Power

Now that the total circuit current has been calculated, the apparent power or volt-amps (VA) can be calculated using the total current in the following formula:

$$VA = E \times I_T$$

$$VA = 220 \times 28.62$$

$$VA = 6296 \text{ VA}$$

We can also use the power triangle to verify the previous values. With watts at 4032 and VARs at 4840 we can simply use the Pythagorean theorem to calculate the hypotenuse also know as the apparent power.

$$VA = \sqrt{W^2 + VARs^2}$$

$$VA = \sqrt{4032^2 + 4840^2}$$

$$VA = 6299.4 \text{ VA}$$

The slight discrepancy in the final VA numbers is due to rounding of numbers.

POWER FACTOR

Since the volt-amps and watts are known, the power factor (PF) can be determined using the following formula:

$$PF = \frac{\text{true power}}{\text{apparent power}}$$

$$\%PF = \left(\frac{W}{VA}\right) \times 100$$

$$\%PF = \left(\frac{4026 \text{ W}}{6296 \text{ VA}}\right) \times 100$$

$$\%PF = .64 \times 100$$

$$PF = 64\%$$

TechTip!

Power Factor can also be calculated using

$$PF = \cos\theta = \frac{Z}{R}$$

ANGLE THETA (θ)

Remember that the power factor is the cosine of angle theta (θ). Therefore, angle θ is:

$$\cos \theta = 0.64$$

$$\text{arc cos } .64 = \theta$$

$$\theta = 50.2°$$

This 50.2° angle is the exact same angle found when vectoring the circuit currents, thus confirming the congruency of the power triangle and current triangles and verifying that the calculations are correct.

TechTip!

Angle Theta can also be calculated using

$$\text{Angle Theta} = \tan\theta = \frac{R}{X}$$

DETERMINING IMPEDANCE USING THE ASSUMED VOLTAGE METHOD

Given the parallel circuit shown in Figure 10–6 and assuming the following values: $X_{L1} = 24 \ \Omega$, $X_{L2} = 16 \ \Omega$, $X_C = 6 \ \Omega$, $R_1 = 6 \ \Omega$, and $R_2 = 12 \ \Omega$, here is how the assumed voltage method works:

FIGURE 10-6 Schematic diagram of circuit with multiple branches.

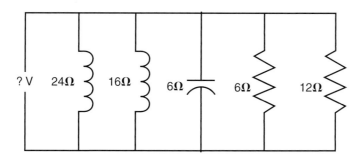

1. When possible, assume a voltage that will result in currents for all parallel legs that are whole numbers. "Assuming" means that we will simply use a voltage value that easily divides by the resistive, inductive reactance, and capacitive reactance values. This assumed (fictitious) voltage is just a temporary value that makes the math easier. When we assume a fictitious voltage, the currents that result from this assumed voltage will be assumed currents. Remember that these currents are not actual measurable currents in the branches.
2. An assumed voltage of 24 V is used for this circuit.
3. Using Ohm's law for AC circuits, the current through each leg will be determined as in Table 10–1:

TABLE 10–1	Table to Show the Individual Branch Circuit Values for Circuit 10–6			
$I_{L1} = \dfrac{E}{X_{L1}}$	$I_{L2} = \dfrac{E}{X_{L2}}$	$I_C = \dfrac{E}{X_C}$	$I_{R1} = \dfrac{E}{R_1}$	$I_{R2} = \dfrac{E}{R_2}$
$I_{L1} = \dfrac{24V}{24}$	$I_{L2} = \dfrac{24V}{16}$	$I_C = \dfrac{24V}{6}$	$I_{R1} = \dfrac{24V}{6}$	$I_{R2} = \dfrac{24V}{12}$
$I_{L1} = 1A$	$I_{L2} = 1.5A$	$I_C = 4A$	$I_{R1} = 4A$	$I_{R2} = 2A$

4. The total current for the circuit can now be determined for the assumed voltage of 24 by the formula, $I_T = \sqrt{I_R^2 + (I_C - I_L)^2}$, where the absolute value of $I_C - I_L$ is used, or by creating a current vector diagram as shown in Figure 10-7:

$$I_T = \sqrt{I_{RT}^2 + (I_C - I_{LT})^2}$$
$$I_T = \sqrt{6\ A^2 + (4\ A - 2.5\ A)^2}$$
$$I_T = \sqrt{6^2 + (1.5)^2}$$
$$I_T = \sqrt{36 + (2.25)}$$
$$I_T = \sqrt{38.25}$$
$$I_T = 6.18\ A$$

FIGURE 10–7   Vector diagrams of branch currents for Figure 10–6.

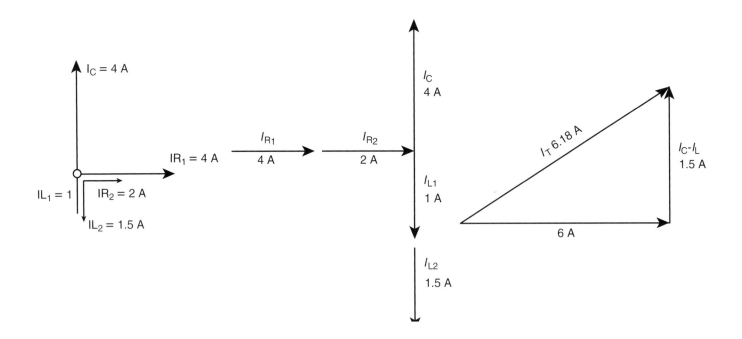

or, by using the current triangle and adding vectors we arrive at the same circuit total current of 6.18 A.

5. With the assumed 24 V power supply applied, the circuit current is 6.18 A. The real impedance of the circuit can now be determined by using Ohm's law:

$$Z_T = \frac{E}{I_T}$$

$$Z_T = \frac{24 \text{ V assumed}}{6.18 \text{ A assumed}}$$

$$Z_T \text{ (actual)} = 3.88 \ \Omega$$

This method is sometimes easier than the reciprocal method for determining the impedance of a parallel circuit. To see how assuming a different voltage supply affects the circuit, substitute 12 V for the assumed voltage and then assume 48 V. You will conclude that no matter which voltage you "assume" the total Z is the same. Remember that the currents you solved for are not actual circuit currents:

$$Z_T @ 12 \text{ V: } Z_T = 3.88 \ \Omega$$

$$Z_T @ 48 \text{ V: } Z_T = 3.88 \ \Omega$$

PARALLEL RESONANT CIRCUITS

You know that in a parallel circuit the inductive current and capacitive current are 180° out of phase and subtract from each other. As you can verify from the vector diagrams of the currents, if the capacitive current equals the inductive current then the net reactive current for the circuit is zero. This produces a minimum line current at the point of resonance. Again, resonance is the frequency point for the circuit where X_L equals X_C and that means that I_L equals I_C. This minimum line current indicates that there must have been maximum impedance. In theory, when a parallel LC circuit reaches resonance, the total circuit current should reach zero. Figure 10–8 shows the current relationship at resonance. Figure 10–9 shows a graph of frequency vs. line current around resonant frequency (f_R). At resonant frequency, the capacitive reactive current and inductive reactance current are equal and cancel.

Tech Tip!

The flywheel effect of a resonant circuit refers to the mechanical flywheel concept. A mechanical flywheel has the effect of using the inertia of initial energy to keep a wheel spinning even during periods when the input energy has disappeared. In the electrical "flywheel" the circuit appears to have energy in the system even though the initial energy from an input source has disappeared. Even when the input sine wave is at zero volts, there is still current flowing in the "LC tank" between the capacitor and the inductor. Therefore, the current continues to flow even though the "flywheel" circuit is at a point with no further input energy.

FIGURE 10–8 Sine waves of current that occur at resonance.

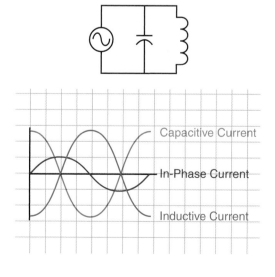

TechTip!

This internal "tank" circuit assumes that there are no losses of energy within the tank. As you know, this is not entirely true or we would have invented perpetual motion. In fact there is resistance in the coil circuit and some resistance in the capacitor circuit. There is resistance to the connecting circuit wiring as well that will produce watts loss as the current moves from capacitor to coil and back, as was discussed in line loss in the DC Theory text. The tank circuit does need energy from the line but the concept of a parallel tank circuit is used extensively in the practice of power factor correction. This common practice of using a tank circuit to correct power factor will be discussed in detail later in a different text where power factor is explained in much greater detail.

Figure 10–10 is an LC resonant circuit called a "tank" circuit. Note that the current in the inductor or capacitor branch of the circuit is very high at resonance when compared to the total line current. The individual branch currents do not affect the I_T line current because each is 90° out of phase with the "in phase" line current and opposite each other. As the AC-supply voltage reverses polarity, the stored energy in the inductor's magnetic field and capacitor's electrostatic field charge and discharge within the "tank". This cycling effect generates a sine wave at the resonant frequency of the LC parallel branches. This sine wave generation is called the flywheel effect of an LC tank circuit.

As you learned earlier, an increase in frequency decreases capacitive reactance, increases total current, and decreases impedance. An increase in frequency also decreases the power factor. See Table 10–2 to see the effects of changing variables in an RLC parallel circuit. As you will note, some of the values in Table 10–2 have question marks as those answers will depend on the original circuit before increasing or decreasing a variable. The top two rows are based on increasing or decreasing frequency from resonance.

FIGURE 10–9 The effects of resonant frequency on a parallel RLC circuit yields minimum line current.

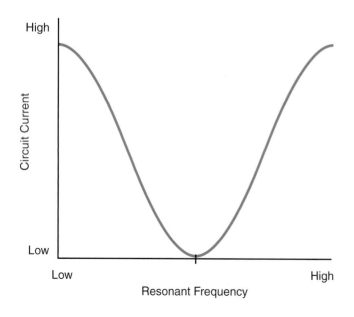

FIGURE 10–10 A "tank" circuit created by L and C in parallel.

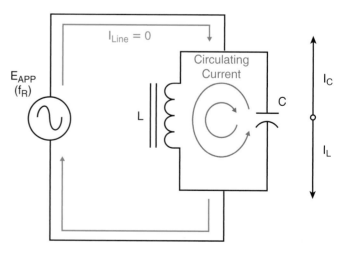

| TABLE 10-2 | | Trends of Increase, Decrease, or Stay the Same for Changing One Variable At a Time | | | | | | | | | | |

Variable	X_C	I_L	X_C	I_C	I_R	I_T	Z_T	VA	P_T	VARs	θ	%PF
Freq Inc Resonance	Inc	Dec	Dec	Inc	Same	Inc	Dec	Inc	Same	Inc	Inc	Dec
Freq Dec Resonance	Dec	Inc	Inc	Dec	Same	Inc	Dec	Inc	Same	Inc	Inc	Dec
Induc Inc	Inc	Dec	Same	Same	Same	?	?	?	Same	Dec	Dec	Inc
Induc Dec	Dec	Inc	Same	Same	Same	?	?	?	Same	Inc	Inc	Dec
Cap Inc	Same	Same	Dec	Inc	Same	?	?	?	Same	Inc	Inc	Dec
Cap Dec	Same	Same	Inc	Dec	Same	?	?	?	Same	Dec	Dec	Inc
Resis Inc	Same	Same	Same	Same	Dec	Inc	Dec	Dec	Dec	Same	Inc	Dec
Resis Dec	Same	Same	Same	Same	Inc	Dec	Inc	Inc	Inc	Same	Dec	Inc
E supply increase	Same	Inc	Same	Inc	Inc	Inc	Same	Inc	Inc	Inc	Same	Same
E supply Decrease	Same	Dec	Same	Dec	Dec	Dec	Same	Dec	Dec	Dec	Same	Same

SUMMARY

The voltage across all legs of a parallel RLC circuit is the same. The current flow in the resistive leg will be in phase with the voltage. The current flow in the inductive leg will lag the voltage by 90° (ELI), and the current flow in the capacitive branch will lead the voltage by 90° (ICE).

The total current can be found by using a vector diagram and trigonometry or by using an algebraic formula (see section 10.8). Remember to find the currents in a parallel circuit instead of vectoring oppositions calculating impedance as in series circuits. The angle found using the current triangle vector is the phase angle that tells us whether the current is leading or lagging the line voltage, and by how many degrees. The angle to the horizontal in the triangle vector is the same angle that we calculate for the circuit power triangle.

The impedance (Z_T) is equal to the applied voltage divided by the total circuit current (I_T). Z_T is maximum at resonance frequency (f_R) because the I_T is minimum. We can also find Z_T by using an assumed voltage that is chosen just to make the math easier. With the assumed voltage we get assumed currents, but the result is that we find the actual circuit impedance, no matter what voltage we assume.

A resonant LC tank circuit is very useful for generating a sine wave at the tank's resonant frequency. The idea of a resonant tank circuit will also be used to explain power factor correction. This is called the flywheel effect.

Now we have a fairly good understanding of an RLC parallel circuit. From this understanding, we can change one variable at a time and predict how that would change the circuit results. Because the reactance of the coils and capacitor changes with frequency, we would have to know the starting point to determine how increasing or decreasing the frequency would alter the entire circuit characteristics.

REVIEW QUESTIONS

1. In a parallel RLC circuit, what is the relationship between the following?
   a. Applied voltage and branch voltage
   b. Total current and individual branch current
   c. Component voltage and component current in R, L, or C component branches
2. In Figure 10–3B, why is the Pythagorean theorem used to calculate the *total* current?
3. Power factor is the ratio of ____ to ____. Explain.
4. What determines the phase angle between the total current and the applied voltage in a parallel RLC circuit?
5. A tank circuit operates on the so-called flywheel effect. Explain.

6. Explain why assuming a voltage will result in calculating the true circuit impedance.
7. If an RLC parallel circuit is at resonance, increasing the frequency will make the circuit appear more _____.
8. What affect does changing the frequency below resonance have on the true power of the circuit?
9. By adding more capacitors to the resonant circuit, the total current of the circuit will _____.
10. At resonance the power factor of the circuit is at _____ %.

PRACTICE PROBLEMS

All the practice problems refer to Figure 10–11. Find the total reactance in the following figure.

1. Find the current in each branch.
2. Find I_T.
3. Find Z_T.
4. Using E_T and I_T, find Z_T. Does this answer agree with Question 3?
5. What is the true power in this circuit?
6. What is the apparent power (VA)?
7. What is the power factor?
8. What is the phase angle?
9. Find the Z_T using the assumed voltage method other than that given. Does it agree with Question 3?
10. Draw the vector diagram for the power triangle for the total circuit.

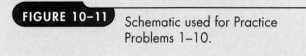

FIGURE 10–11 Schematic used for Practice Problems 1–10.

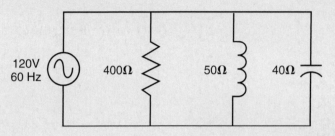

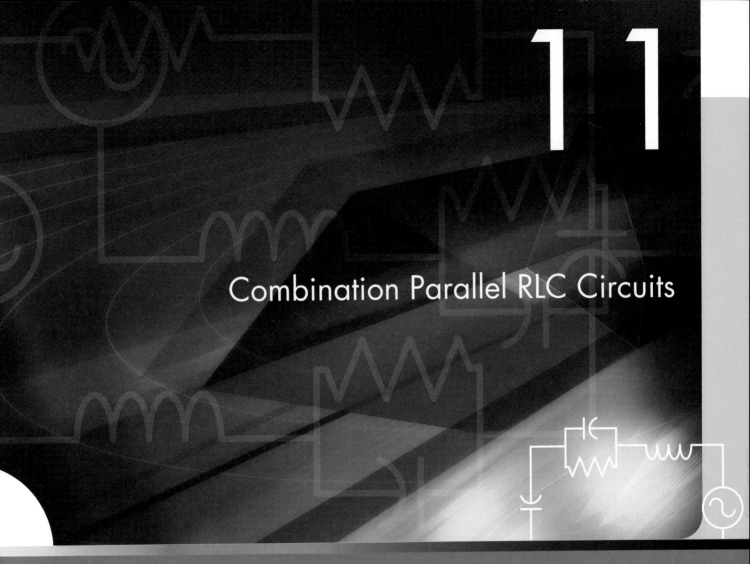

11

Combination Parallel RLC Circuits

O U T L I N E

OVERVIEW

Most AC circuits have some combination of resistance, capacitance, and inductance. In fact, many electrical systems have series and parallel combinations of components in the same circuit. These types of circuits are referred to as combination circuits or as complex circuits. Because of the different phase angles involved, vector analysis is the only sure way to analyze a combination circuit of this type.

This chapter teaches you how to use your vector analysis skills in the solution of combination circuits. Remember that all the information you learned previously still applies. Series circuits rules apply to series circuits and parallel rules to parallel circuits at the circuit segment level. Tools such as superposition, Kirchhoff's laws, Thevenin equivalent circuits, and other such methods will work in AC combination circuits. However, the calculations must be made using vector analysis. In this chapter we will examine series circuits in parallel with other series circuits and parallel circuits in series with other parallel circuits.

Power factor correction is explained and the needed calculations for single phase circuits will be explored. We will also approach the troubleshooting concept of taking readings and calculating back into the circuit what unknown components must have been involved; a "forensic" analysis.

OBJECTIVES

After completing this chapter, you should be able to:
- Solve for all quantities in a combination RLC AC circuit
- Be able to draw vectors for current, impedance, voltage, and triangles for impedance and power as dictated by the circuit type
- Determine a lead or lag phase angle at any point in a complex circuit
- Solve for power factor and be able to correct a lagging power factor to a predetermined percentage or to unity
- Find unknown component values from actual readings in a circuit

COMBINATION RLC CIRCUITS

Using the current for the vector reference in a series circuit simplifies the analysis. Conversely, the voltage reference is the simplest choice for a parallel circuit. The secret for solving for unknowns in any combination RLC circuit is to reduce the complex circuit to a single equivalent load and source for voltage. This method was taught in the DC Theory text and is called the Thevenin equivalent circuit. One of the differences between DC and AC is that the equivalent *resistance* in DC must be called equivalent *impedance* in AC, and the Thevenin voltage may have a phase angle associated with it.

This approach is similar to the approach used in solving for total resistance in series/parallel combinations in DC Theory. The difference is that in an AC circuit containing inductance, capacitance, and resistance, the vector components of the series equivalent circuit will not be in phase with each other. These series vectors must be reduced to their vertical and horizontal components, which can then be vectorially added together. The sum of the vertical components can be added to the sum of the horizontal components using the Pythagorean theorem.

CIRCUIT ANALYSIS: RESISTANCE, INDUCTANCE, AND CAPACITANCE

Consider the procedure to solve for the parameters of combination circuits containing inductance, capacitance, and resistance. Take a look at the circuit in Figure 11–1. The circuit is largely a series circuit with two "zones" of parallel circuits. Each component or parallel combination is identified as vector impedance. For example, the 10 Ω series resistor is Z_1, the combination of the 8 Ω resistor and 8 Ω X_L inductor in parallel is Z_2, and so forth around the circuit. As with any complex circuit we need to break the small zone circuits down to manageable equivalent components. Because the large circuit can be identified as a series, we start by creating a series component equivalent of each of the parallel sections.

FieldNote!

There is tendency to think that many of the tough circuit evaluations are not used in the electrician's everyday work. Even though you don't often see an electrician with a calculator in his tool pouch, the knowledge is critical to understand circuits when they don't seem to be working properly, or the values of the circuit measurement don't appear to be correct. An electrician, who has a working knowledge of the complex circuits and the characteristics that can seem improbable, will be better qualified to diagnose circuit problems.

FIGURE 11–1 Schematic of combination RLC circuit.

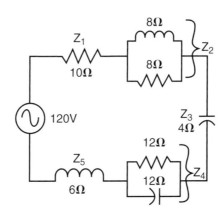

11.1 Parallel Component Vector Addition

In order to add these vector components together, the magnitude and direction of the parallel combinations (Z_2 and Z_4) must first be solved. To determine the impedance of each of these vectors, use the assumed voltage method as discussed in Chapter 10.

We do not know the actual voltage across the parallel branch yet but we can still find the parallel circuit impedance by vectoring the assumed currents from an assumed voltage. For Z_2, assume a voltage of 8 V across the parallel components. Applying the assumed voltage of 8 V, an assumed 1 A will flow through each component. The current through the resistor will be in phase with the voltage, and the current through the inductor will lag the voltage by 90°:

$$I_{Z_2} = \sqrt{I_L{}^2 + I_R{}^2}$$

$$I_{Z_2} = \sqrt{1 + 1}$$

Using the Pythagorean theorem, the total current equals the square root of 2 as below:

$$I_{Z_2} = \sqrt{2}$$

$$I_{Z_2} = 1.414 \text{ A @ } 45° \text{ lag}$$

$$Z_2 = \frac{E_{\text{assumed}}}{I_{Z_2}}$$

$$Z_2 = \frac{8 \text{ V}}{1.414 \text{ A}}$$

$$Z_2 = 5.66 \ \Omega \text{ @ } 45° \text{ lag}$$

Therefore, Z_2 has an impedance of 5.66 Ω at 45° lag.

Of course, this same result could be reached quickly with a scientific calculator as follows. For Z_2, you know that:

$$\overline{Z}_2 = \frac{\overline{X}_L \times \overline{R}}{\overline{X}_L + \overline{R}}$$

where the bar over the Z, X_L, and R means that they are vectors. You also know that:

$$\overline{X}_L = 8 \ \angle 90° \text{ and } \overline{R} = 8 \ \angle 0°$$

substituting:

$$\overline{Z}_2 = \frac{(8\angle 90° \times 8\angle 0°)}{(8\angle 90° + 8\angle 0°)}$$

Performing this vector calculation with a scientific calculator yields:

$$\overline{Z}_2 = 5.66 \ \Omega \text{ @} 45° \text{ lag}$$

Repeat the same process for Z_4, using the assumed voltage method or a scientific calculator to find the Z_4 impedance. Using 12 V as the assumed voltage gives a current through Z_4 of 1.414 A, the same as for Z_2 (you can see how assuming a "workable" voltage can make this portion of the problem solving easy).

TechTip!

The tangent of θ_{Z_2} is equal to:

$$\tan \theta_{Z_2} = \frac{I_L}{I_R}$$

$$\tan \theta_{Z_2} = \frac{1 \text{ A}}{1 \text{ A}}$$

$$\tan \theta_{Z_2} = 1$$

$$\theta_{Z_2} = 45°$$

FIGURE 11–1

Schematic of combination RLC circuit.

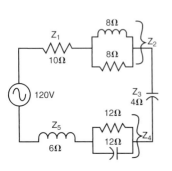

Using the assumed voltage and newly found current, solve for the impedance of:

$$Z_4 = \frac{E \text{ assumed}}{I \text{ assumed}}$$

$$I_{Z_4} = \frac{12 \text{ V}}{1.414 \text{ A}}$$

$$Z_4 = 8.49 \ \Omega \ @ \ 45° \text{ lead}$$

Since the value of the current through the resistor and capacitor are both 1 A, the tan θ will be same as the tan θ through the Z_2 group because the current flow through those components was also equal to 1 A. The tangent of that angle is also equal to 1. Since this is a capacitive circuit, the angle will be negative instead of positive; therefore, θ = −45°, and $Z_2 = 8.49 \ \Omega \ @-45°$ or leading effect.

If the capacitor and inductor had different values relative to the resistor in the group, the angle theta (θ) would calculate to a different magnitude. Because our current through each of the components is equal (the resistance and reactance being equal), the current flow through each component is also equal in our example circuit.

Just for sake of clarity, we can start a chart or table to keep track on the individual circuit "zones" as depicted in Table 11–1.

TABLE 11–1   Table to Track Each "Zone" Impedance for Figure 11–1

	Z_1	Z_2	Z_3	Z_4	Z_5
I	?	Assumed 1.414 A	?	Assumed 1.414 A	?
Z	10Ω @ 0°	Actual 5.66Ω @ 45° lag	4Ω @ 90° lead	Actual 8.49Ω @ 45° lead	6Ω @ 90° lag

Here is another method. Both Z_2 and Z_2 combinations could be separately solved using the formula but without the vector direction reference.

$$Z = \frac{R \times X}{\sqrt{R^2 + X^2}}$$

For example, using Z_2 parameters:

$$Z_2 = \frac{8 \times 8}{\sqrt{8^2 + 8^2}}$$

$$Z_2 = \frac{64}{11.31} = 5.66 \ \Omega$$

CIRCUIT TOTAL VECTOR ADDITION

To find the circuit impedance we vectorially add the oppositions in series. Figure 11–2 shows the five impedance values of Figure 11–1 added together graphically using the arrowhead-to-end or head-to-tail method. The resultant vector for this circuit is 20 Ω at 0°.

FIGURE 11–1

Schematic of combination RLC circuit.

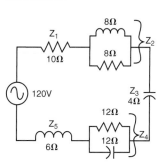

FIGURE 11-2 Vector addition "arrowhead method" for series circuit impedance.

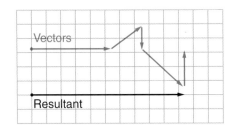

Note that Z_3 and Z_5 have no horizontal component since they are a pure capacitance and pure inductance, respectively. Likewise, Z_1 has no vertical component since it is a pure resistance.

This is the solution for this problem. Vectors Z_2 and Z_4 must be reduced to their vertical and horizontal components. Table 11–2 lists the vertical and horizontal components and their calculation. The information in this table is based on vector addition of horizontal and vertical components found using trigonometry. The vector addition as seen in Figure 11–3 is the result of breaking each vector into its horizontal and vertical components.

Note that the negative values for Z_3 and Z_4 are due to the capacitive lag of the parallel components.

TABLE 11-2 Table to Find Horizontal and Vertical Components of Vectors for Vector Addition

Component	Z_1 (Ω)	Z_2 (Ω)	Z_3 (Ω)	Z_4 (Ω)	Z_5 (Ω)
Vertical	0	$Z_2 \times \sin(45°) = 4\ \Omega$	-4	$Z_4 \times \sin(-45°) = -6$	6
Horizontal	10	$Z_2 \times \sin(45°) = 4$	0	$Z_4 \times \cos(-45°) = 6$	0

The net resultant of the total opposition provided by this circuit impedance triangle is 20 ohms of impedance at 0° phase angle.

FIGURE 11-1

Schematic of combination RLC circuit.

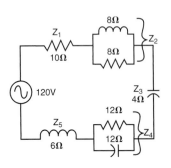

VOLTAGE DROP CALCULATIONS FOR COMBINATION CIRCUITS

Now that we know the circuit impedance of all the components combined into one equivalent component, we can now determine the circuit current when 120 V AC is applied. The simple Ohm's law formula of $I = \dfrac{E}{Z}$ will tell us that the circuit current is 6 A and the phase angle is 0° between the line current and the line voltage (they are in-phase with each other). If we know the current is 6 A, and the current is the same through the series circuit, we can determine the volt drop on each "zone" as described earlier (Table 11–3).

Table 11–3 summarizes the results.

FIGURE 11-3   The opposition provided by the components is represented by horizontal and vertical components.

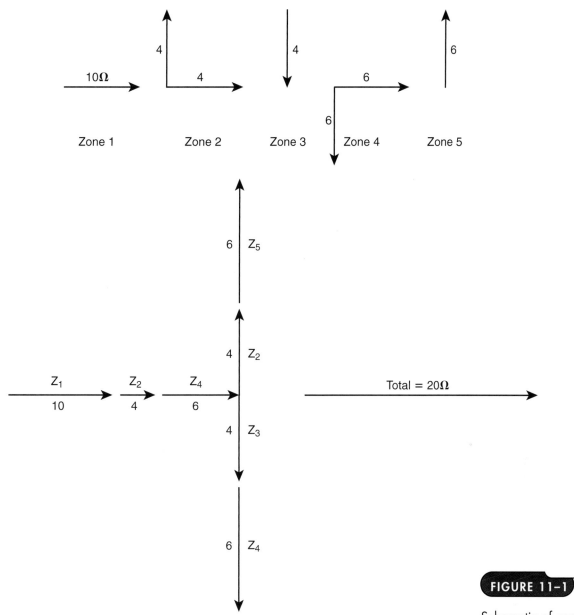

Zone 1   Zone 2   Zone 3   Zone 4   Zone 5

Total = 20Ω

FIGURE 11-1

Schematic of combination RLC circuit.

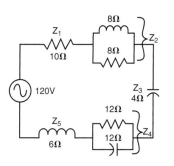

TABLE 11-3   Table to Calculate the Voltage Drop of Each "Zone" of Figure 11–1

Z_1	Z_2	Z_3	Z_4	Z_5
$E_{Z1} = I_T \times Z_1$	$E_{Z2} = I_T \times Z_2$	$E_{Z3} = I_T \times Z_3$	$E_{Z4} = I_T \times Z_4$	$E_{Z5} = I_T \times Z_5$
$E_{Z1} = 6A \times$ 10 Ω	$E_{Z2} = 6A \times$ 5.66Ω	$E_{Z3} = 6 A \times$ 4 Ω	$E_{Z4} = 6 A \times$ 8.49 Ω	$E_{Z5} = 6 A \times$ 6 Ω
$E_{Z1} = 60$ V @ 0° resis	$E_{Z2} = 33.96$ V @45° ind	$E_{Z3} = 24$ V @90° cap	$E_{Z4} = 50.94$ V @45° cap	$E_{Z5} = 36$ V @90° ind

Figure 11–4 shows the vector addition of the voltages. As you note the voltages added arithmetically add up to much more than the original applied voltage of 120 V. Adding these voltage vectors with the known phase angles will again produce a vector sum of 120 V AC. These are actual voltages that can be measured across each of the zone components. One of the methods to find the parallel impedance of a zone used an assumed voltage. This is *not* an assumed voltage but the actual voltage of the circuit. From this real voltage we can now find the real current in each of the parallel component zones.

FIGURE 11–4 Voltage vector addition of sereies voltage drops for Circuit 11–1.

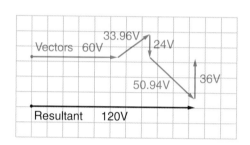

For Z_2, each of the two components has an ohmic value of 8 Ω with 33.96 V across them. Each 8 Ω component of the parallel will have current through it as determined by $I = \dfrac{E}{ohms}$. Using Ohm's law we calculate 4.245 A of current through the resistance leg, and 4.245 A of lagging current through the inductive leg, lagging by 90°. Remember that for parallel circuits we add currents, not oppositions. As we add these two current that are 90° out of phase with each other we will again have 6 A as we expected from the line current calculation. We know this 6 A is 45° out of phase with the voltage across the zone 2 components. Using the 6 A and the 33.96 V we confirm the real impedance is $Z = \dfrac{E}{I}$ or $\dfrac{33.96\text{ V}}{6\text{ A}} = 5.66$ Ω of Z.

For Z_4, each of the two components has an ohm value of 12 Ω with 50.94 V across them. Each 12-Ω leg will have 4.245 A. The current through the resistive leg will lag the current in the capacitive leg by 90°. Again by adding these current vectors, we return to the line current of 6 A. The actual impedance of this zone 4 is: $Z = \dfrac{E}{I}$ or $\dfrac{50.94\text{ V}}{6\text{ A}} = 8.49$ Ω of Z, confirming our original assumed voltage method of finding the Z of the parallel zone.

POWER CALCULATIONS

The power calculations will follow a simple format of using the current through each component and the opposition of each component. The formula of: $I^2R = P$, using X_L or X_C for R, will allow us to determine the power whether true or reactive for each component. Table 11–4 tracks the calculation for each zone.

FIGURE 11–1

Schematic of combination RLC circuit.

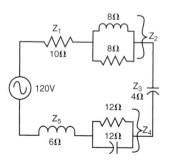

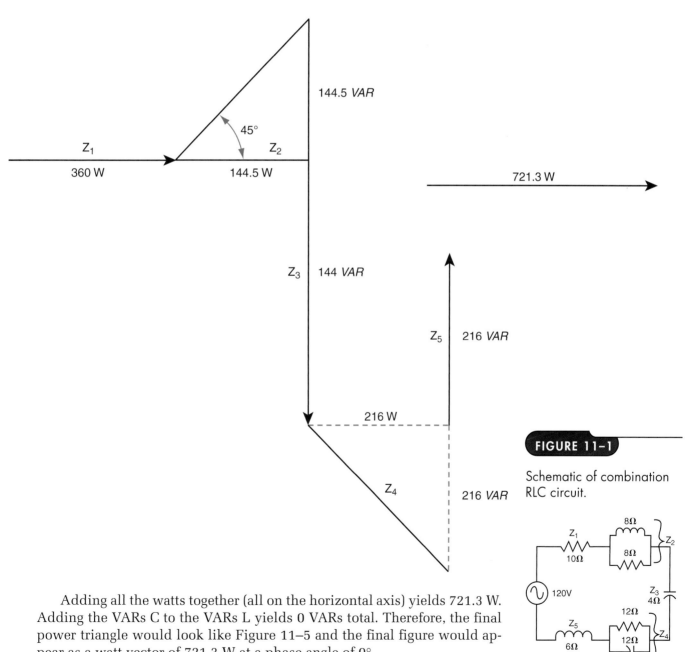

TABLE 11-4	Table to Calculate Each Zone Power Calculations for Circuit 11–1			
Z_1	Z_2	Z_3	Z_4	Z_5
$6^2 \times 10\ \Omega$	$4.25^2 \times 8\ \Omega$ $4.25^2 \times 8\ \Omega$	$6^2 \times 4\ \Omega$	$4.25^2 \times 12\ \Omega$ $4.25^2 \times 12\ \Omega$	$6^2 \times 6\ \Omega$
360 W	144.5 W 144.5 VARs L	144 VARs C	216.8 W 216.8 VARs C	216 VARs L

FIGURE 11-5   Final power triangle for power calculations used for Figure 11–1.

FIGURE 11-1

Schematic of combination RLC circuit.

Adding all the watts together (all on the horizontal axis) yields 721.3 W. Adding the VARs C to the VARs L yields 0 VARs total. Therefore, the final power triangle would look like Figure 11–5 and the final figure would appear as a watt vector of 721.3 W at a phase angle of 0°.

This should be the exact same as the apparent power of 6 A × 120 V or 720 VA. Because of rounding numbers, these two calculations are not exactly the same. However, they are close enough to confirm the procedure. We can also confirm that the phase angle of the power triangle is 0° as was the original impedance triangle. $\dfrac{\text{Watts}}{\text{VA}}$ also yields 1 or **unity power factor** known as 100%. Power factor expressed as a percent of the apparent power that is consumed as true power is also defined as the cosine θ × 100 = %PF. Cos 0° is 1 × 100 = 100%.

Unity power factor
Ratio of true power (watts) divided by apparent power (volt-amps) with the dividend as *one,* or the ratio is 1. Unity power factor is also known as 100%, meaning that 100% of the volt-amps are used as watts and the reactive power does not affect the circuit. Power factor can not be over unity or 100%.

SERIES CIRCUITS IN PARALLEL

The other variation of a combination circuit is the series circuits in parallel with each other. Consider Figure 11–6. This circuit is very typical in a small commercial application where there are a number of different motors on a branch circuit, all operating at one time. Each of the series circuits is typical of a motor that has inductance and resistance in series. We will want to correct the power factor created by all the inductive loads of the motors so we will find the value of the last parallel component—a capacitor.

FIGURE 11–6   Typical schematic of series circuits in parallel.

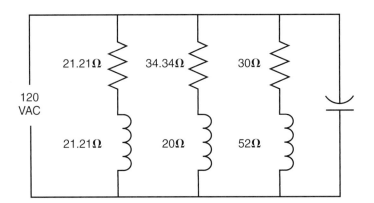

Note: the Capacitor value is to be determined

FIGURE 11–1

Schematic of combination RLC circuit.

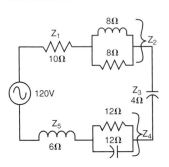

To analyze the circuit, first break the circuit down into smaller segments to be treated individually then assembled into the circuit equivalent component. Branch 1 ("impedance Z_2") is a series circuit with 21.21 Ω of R and 21.21 Ω of X_L. Branch 2 (Z_2) is a motor with 34.34 Ω of R and 20 Ω of inductive reactance. Branch 3 (Z_3) is a motor with 30 Ω of R and 52 Ω of X_L. Branch 4 (Z_4) is a capacitor. Build a table as shown in Table 11–5 to help track the branches values. Construct a Z triangle for each series circuit to find the impedance of each branch to fill in the column for impedance. Then, calculate the current in each branch based on the impedance.

Because the branches are all in parallel the next step is to vector the currents calculated in Table 11–5. The vector diagram in Figure 11–7 represents this vector addition.

TABLE 11-5	A Table to Track the Z and I or Each Branch of Parallel-Series Circuits

$Z_1 = \sqrt{R^2 + X_L^2}$	$Z_2 = \sqrt{R^2 + X_L^2}$	$Z_3 = \sqrt{R^2 + X_L^2}$	Z_4
$Z = \sqrt{21.21^2 + 21.21^2}$	$Z = \sqrt{34.34^2 + 20^2}$	$Z = \sqrt{30^2 + 52^2}$	?
$Z = 30\ \Omega\ @\ 45°\ \text{lag}$	$40\ \Omega\ @\ 30°\ \text{lag}$	$60\ \Omega\ @\ 60°\ \text{lag}$	
$I_1 = \dfrac{120\ V}{30\ \Omega} = 4\ A$	$I_2 = \dfrac{120\ V}{40\ \Omega} = 3\ A$	$I_3 = \dfrac{120\ V}{60\ \Omega} = 2\ A$	
@ 45° lag	@ 30° lag	@ 60° lag	

FIGURE 11-7	Using the horizontal and vertical components or each branch current vector will create a resultant current for the first three branches of Figure 11-6.

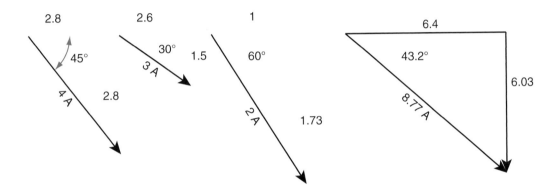

You may find the vertical and horizontal components of each branch current vector to add them, or use a scientific calculator to find the vector sum. As you see in the vector diagram the horizontal components all add to yield 6.4 A and the vertical components add to yield approximately 6 A for a vector sum of 8.77 A at phase angle to the line voltage of 43.2° lag. This angle tells us that the power factor so far is the cosine of $\theta \times 100$ or 72.9%.

The voltage across each branch of the parallel circuit is the line voltage of 120 V. Even though the components are depicted as separate R and X_L components, they are the same motor component containing resistance and cannot be measured individually as E_{XL} and E_R, but can be measured as a total component voltage.

POWER FACTOR CORRECTION

As we have seen in the previous circuit in Figure 11-6 and Figure 11-7 we have a power factor of 72.9% lag without a capacitor in the circuit yet. This power factor is a result of the inductive current causing reactive power that is held in the magnetic field of the coils every time the magnetic field expands and collapses.

TechTip!

FieldNote!

This effect creates a system inefficiency, which means we must deliver more current to the circuit than is actually needed to support the true power—or real work. If we can reduce the apparent power to the circuit, in this case 8.77 A × 120 V = 1052.4 VA, then the line current can be reduced and less current is needed from the source. The solution to drawing additional current from the line is to create a "tank" circuit using a capacitor as one-half of the tank—as discussed in Chapter 10.

If we want to eliminate all excess current from the line, we will need to supply all the inductive current from the "tank" of the capacitive current. In other words we will offset the 6 A of inductive current with 6 A of capacitive current. Working backwards from what we need to supply, we can find the amount of needed X_C to create 6 A of I_C at 120 V: $X_C = \dfrac{E}{I_C X_C} = \dfrac{120\,\text{V}}{6\,\text{A}} = 20\,\Omega X_C$.

From this information we can now calculate C in farads by the formula $C = \dfrac{1}{2\pi f\,C}; C = \dfrac{1}{(377)\,(20)} = .0001326$ farads $= 132.6\,\mu\text{F}$ (when 60 Hz is applied). Therefore, the needed capacitance to correct the power factor to 100% is 132.6 µF.

Suppose a 43.2° phase shift is too large but we don't want to correct to exactly 100%. To do this we only need to determine how far we do want to correct. Let us correct to a standard figure of 95% power factor (PF). To do this we need to see what the current triangle would look like if the PF were 95%. The angle is found by using the arc-cos θ = .95. The angle θ is now established at 18.2°, instead of the original 42.3°. If we know the angle and the horizontal vector of resistive current stays at 6.4 A, then the vertical component can be calculated. Using trigonometry, the vertical side or remaining reactive current after correction will be tan 18.2° = (opposite over adjacent of 6.4 A). The opposite side is 2.1 A remaining after correction from the capacitor. The uncorrected current due to inductors was 6 A and 2 A remains; therefore we added 4 A of capacitive current to the circuit by adding a capacitor in parallel. Use the formulas below to determine the capacitor value:

$$X_C = \frac{E}{I_C}$$

$$X_C = \frac{120\,\text{V}}{6\text{A}}$$

$$X_C = 30\,\Omega$$

Then:

$$C = \frac{1}{2\,\pi f\,X_C}$$

Note: C is in Farads.

$$C = \frac{1}{(377)\,30}$$

$$C = 88.4\,\mu\text{F}$$

As you can see, we added a smaller microfarad capacitor to correct the power factor to 95% than we needed to correct the power factor to 100%.

COMPLEX CIRCUIT EVALUATION

The last circuit configuration is to create a series circuit with two parallel combinations as in Figure 11–8. We will use parts of the previous parallel circuit to simplify the calculation but reconfigure the circuit to make sure we apply the proper rules. Start by combining small series or parallel circuits into equivalent circuits and record values in a table for easy reference as shown in Table 11–6. Assume a voltage of 120 V across each of the parallel branches.

FIGURE 11–8   Schematic of two parallel circuits in series.

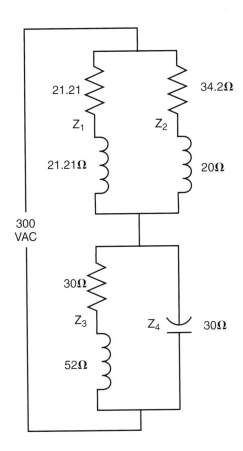

TABLE 11–6   A Table of Each Zone Impedance Using an Assumed Voltage and Current

Z_1	Z_2	Z_3	Z_4
$Z = \sqrt{21.21^2 + 21.21^2}$	$Z = \sqrt{34.34^2 + 20^2}$	$Z = \sqrt{30^2 + 52^2}$	$Z = 30\ \Omega$
$Z = 30\ \Omega$ @ 45° lag	40 Ω @ 30° lag	60 Ω @ 60° lag	30 Ω @ 90° lead
$I_1 = \dfrac{120\ V}{30\ \Omega} = 4\ A$	$I_2 = \dfrac{120\ V}{40\ \Omega} = 3\ A$	$I_3 = \dfrac{120\ V}{60\ \Omega} = 2\ A$	$I = 4\ A$
@ 45° lag	@ 30° lag	@ 60° lag	@ 90° lead

11.2 Impedance Calculation

Because zone 1 and 2 are in parallel, parallel circuit rules apply. Vector the currents, or use algebraic formulas, to determine what actual impedance would be, that would produce this assumed current with the assumed voltage. This calculated value is the actual impedance of Z_1 and Z_2 in parallel. Do the same for Z_3 and Z_4 parallel circuit. Figure 11–9A shows the vector addition for Z_1 and Z_2, and Figure 11–9B shows the vector addition for Z_3 and Z_4. By referring to the figures (A & B) we know the assumed current vector sum if 120 V is assumed as the branch voltage.

FIGURE 11–9 (a) Vector addition of assumed currents in parallel for branches Z_1 and Z_2. (b) Vector addition of assumed currents in parallel for branches Z_3 and Z_4.

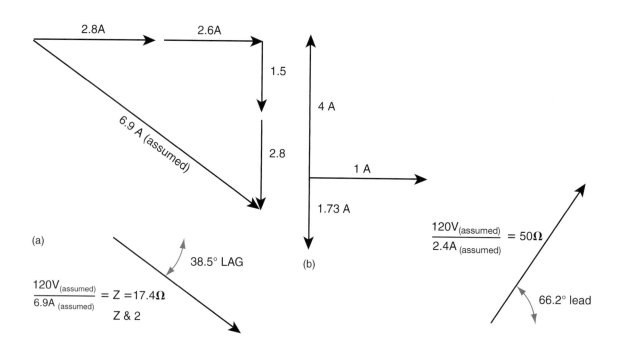

From this we calculate the actual zone impedance of 1–2 as $Z = \dfrac{120\ V_{\text{Assumed}}}{6.9\ A_{\text{Assumed}}} = 17.4\Omega$ of Z at 38.5° inductive effect. Calculate the equivalent impedance of combined Z_3 and Z_4 as $Z = \dfrac{120\ V_{\text{Assumed}}}{2.4\ A_{\text{Assumed}}} = 50\ \Omega$ of Z at an angle of 66.2° capacitive effect. Now the combined equivalent zone 1–2 are in series with the combined zone 3–4 and series circuit rules now apply for adding series impedances. Series circuit rules dictate that we add oppositions to find the total Z of the circuit. Figure 11–10 adds these two equivalent impedances to get a total Z for the circuit of 48.6 Ω at 45.9° capacitive, or a leading-phase angle. With 300 V applied the line current is approximately 6.17 A and it is leading the line voltage by 45.9°.

FIGURE 11–10   Adding the horizontal and vertical components together to yield a circuit impedance triangle.

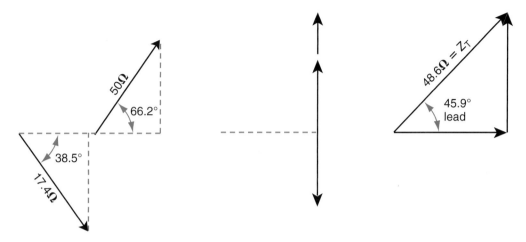

11.3 Voltage Calculation

Now that we have actual circuit impedances and real line current we can determine what voltage drop each component will have as voltage drops. Use the line current as the constant as it is the same through the series circuit in whole. The formula to find the voltage drop across Z_1 and Z_2 is $E = I_T \times Z_{Equiv}$ or $E = 6.17 \times 17.4\ \Omega$ equivalent Z_1 and Z_2 for a volt drop of 107.4 V across that parallel combination. That voltage is leading the circuit current by 38.5° (the current is lagging). The second parallel combination uses $E = I_T \times Z_{Equiv}$ or $6.17 \times 50\ \Omega$ equivalent Z_3 and Z_4 for a drop of 308.4 V. The voltage in this part of the circuit lags behind the line current by 66.2° (current is leading). These two voltages are what we would measure across each of the branches with a voltmeter. We cannot tell with a voltmeter what the phase angle is in reference to the current. We cannot just arithmetically add the voltages because they are at different angles to the line current. By vector addition we find that the total (resultant) line voltage is approximately 300 V, the same as the applied voltage. Again, because of decimal rounding and trigonometric rounding the resultant is not exact. See Figure 11–11 for voltage vector addition using horizontal and vertical components.

FIGURE 11–11   Voltage vector addition using horizontal and vertical components.

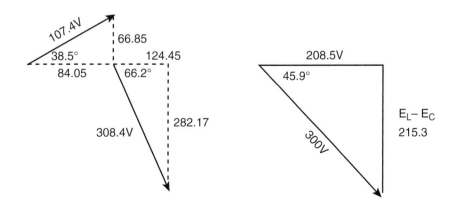

11.4 Power Calculations

The simplest way to calculate the true, apparent, and reactive power is to go to the final voltage triangle vector, and multiply each voltage vector times the line current (Figure 11–12).

FIGURE 11–12 **FIGURE 11–12**   Power calculation triangle for circuit as in Figure 11–8.

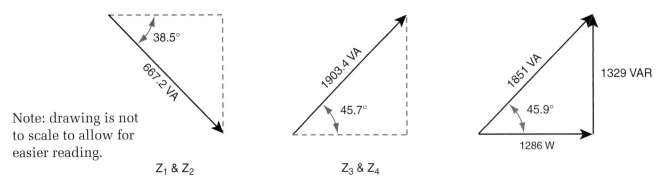

Note: drawing is not to scale to allow for easier reading.

Z_1 & Z_2 Z_3 & Z_4

Zones 1 and 2 have 107.4 V × 6.17 A for 662.7 VA and zones 3 and 4 have 308.5 V × 6.17 A = 1903.4 VA. Total watts for the circuit is 208.5 V @ 0° × 6.17 A = 1286.4 W and reactive VARs: 215.4 V @ 90° × 6.17 A = 1329.0 VARs. The result of all this is apparent power of 300 V @ 45.9° leading × 6.17 A = 1851 VA. The percent power factor is $\left(\dfrac{\text{watts}}{\text{VA}}\right) \times 100$ or $\left(\dfrac{1286.4 \text{ W}}{1851 \text{ VA}}\right) \times 100 = 69.5\%$. The cosine of angle θ (45.9°) × 100 = 69.6 %.

A FORENSIC EVALUATION

We will use a simple parallel circuit to find an unknown value in the circuit based on the reading we can make in the circuit. If we have a simple RLC parallel circuit as in Figure 11–13, and we have reading with an ammeter and a voltmeter and power factor meter, we can decipher the unknown quantities.

FIGURE 11–13   Schematic of RLC circuit with some measurements and an unknown component.

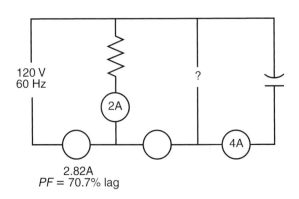

We know that branch 1 is resistive, branch 2 is unknown, and branch 3 is capacitive. The current readings are as shown in Figure 11–12, and the power factor reads 70.7% lagging. We need to find the component of branch 2. Because we have a parallel circuit we know that the currents are added vectorially to get a resultant I_T.

Figure 11–14 shows the current vectors needed to obtain the circuit meter readings. We know that we must have had 6 A of I_L to end with a resultant of 2.82 at a phase angle of 45° (arc-cos .707 = 45°) and the line current is lagging, as it would in a more inductive circuit. Now we can calculate the unknown component as an inductor with a current of 6 A I_L. Since $\frac{E}{I_L} = X_L$ of $\frac{120}{6} = 20\ \Omega$ and $X_L = 2\ \pi fL$ or $L = \frac{X_L}{2\ \pi f}$ then $L = \frac{20\ \Omega}{377}$ or

$L = .053$ H (also known as 53 millihenries). Branch 1 is $\frac{120\ V}{2\ A} = 60\ \Omega$ of resistance. Branch 3 is $\frac{E}{I_C} = X_C$ or $\frac{120\ V}{4\ A} = 30\ \Omega\ X_C$. Then $C = \frac{1}{2\ \pi f\ X_C}$ or

$\frac{1}{(377)\ (30)} = 88.4\ \mu F.$

FIGURE 11–14   Addition of horizontal and vertical components for current vectors of known meter readings, to determine unknown values.

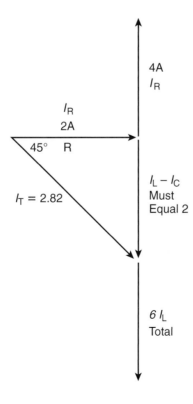

SUMMARY

The combination of series and parallel rules for solving RLC circuits can be confusing. A basic rule is to add oppositions in series and add current in parallel. However, we need to track the actual angles involved as we add the vectors representing current or impedance. As we solve for values in the systems, make sure you note the inductive or capacitive effects of the components involved. It is easy to forget the small details as you solve a large complex problem. Power can be calculated by several means as you use E and I for individual components or use $I^2 \times$ opposition. Then make sure

you note whether it is true, reactive, or apparent power you are calculating.

Using what you know about the circuit characteristics you should be able to calculate any required values in the circuit and also be able to determine what must have been the case as you find readings that indicate how the circuit actually did work. This is a helpful troubleshooting mechanism when you may encounter strange voltages that are only explained by your knowledge of the circuit.

REVIEW QUESTIONS

1. What is the major distinction of a combination or complex circuit?

2. Explain the rules for finding impedance of a series RLC circuit.

3. Explain the rules for finding impedance of a parallel RLC circuit.

4. Explain how to add vectors with different angles and different magnitude.

5. In what type of circuit is current the constant when vectoring voltage components?

6. To correct a lagging power factor and reduce the line current, you need to add a capacitor in _____.

7. Inductive and capacitive reactance cancel each other in an RLC _____ circuit.

8. Unity power factor is also known as _____ percent.

9. If you over correct a lagging power factor, the percent power factor goes _____ and the line current goes _____.

10. Explain how to find the microfarad value of a capacitor if you only know the current to the capacitor and its voltage.

11. In Figure 11–2, vector Z_2 goes up and vector Z_4 goes down. Why?

PRACTICE PROBLEMS

Problems 1 to 4 refer to Figure 11–15.

1. Find $Z_{R2 + L}$ of $L_1 + R_2$ using the assumed voltage method.

2. Find $Z_{R3 + C}$ of $C_1 + R_3$ by the same method as Problem 1.

3. Find Z_{total}.

4. Diagram the circuit (vectorially).

For Questions 5-10 apply 200 V AC at the source.

5. If 200 V AC is applied, find the line current.

6. The voltage drop across R_1 would be _____ with a 200 V AC source voltage.

7. Calculate the phase angle of lead or lag for the total circuit voltage and current.

8. If the source is operating at 60 Hz, find the henry value of L_2.

9. How much voltage would be measured across C_2?

10. Determine the apparent power of this circuit.

FIGURE 11–15 Schematic diagram used for Problems 1–10

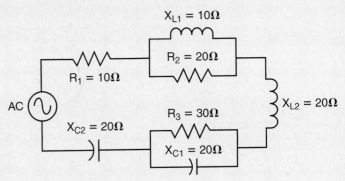

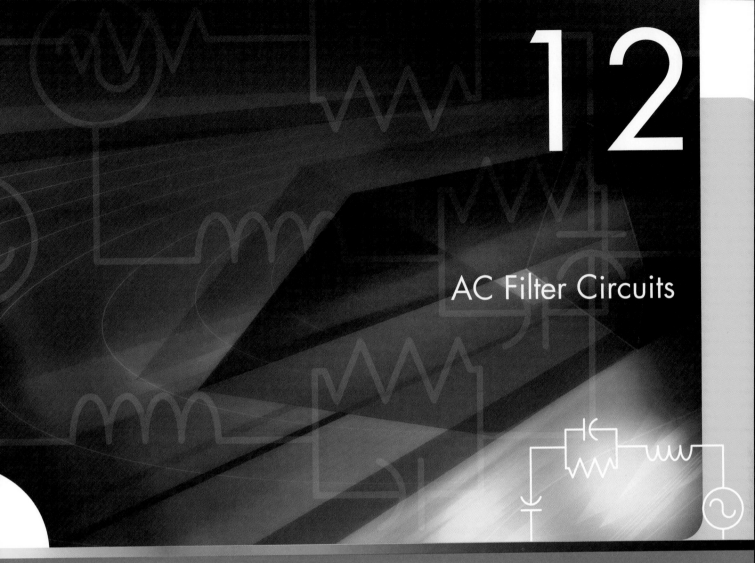

12

AC Filter Circuits

O U T L I N E

OVERVIEW

In previous chapters, you learned about resonance, inductive reactance, and capacitive reactance. These circuit characteristics are used to design useful circuits that will allow current or voltage at certain frequencies to pass to the load, while at other frequencies, current and voltage are filtered out and do not appear at the load. These special circuits are called filters or wave traps.

It is the frequency filter that distinguishes one radio or television station's signal from another by allowing certain frequencies to reach the amplifier circuits while others are trapped at the filter. It is also the filter that prevents one radio station from interfering with another. It is not unusual to have unwanted frequencies generated on an AC power line because of operating various types of loads. These generated frequencies may have considerable power. These same frequencies can cause interference with radios, televisions, computers, or other electronic devices that are operating on the same power line. The most common types of interference are static or a low hum. Power line filters can be installed to prevent these types of interference.

Direct current combined with alternating current is a common occurrence with many electronic circuits. This concept is used in many systems as a base voltage and a signal voltage. The AC signals are sometimes referred to as AC riding on a DC level. This principle is used in circuits that have both DC levels of voltage and superimposed on that DC voltage is an alternating voltage. This is very common in amplifier circuits as the DC level is used to operate the amplifier components such as transistors, and the AC voltage is the signal we want to amplify. The DC level is a component that is not to be passed on from stage to stage of an amplifier, while the AC component is to be passed on.

In this chapter, you will learn about the various types of filters and how they work. This information is used in the design of a wide variety of filters ranging from radio and television circuitry to harmonic filters used in a broad range of power system applications.

OBJECTIVES

After completing this chapter, you should be able to:
- Explain the theory of operation of common type filters
- Identify low-pass filters
- Explain why half-power point is significant
- Calculate values for high-pass filters
- Draw the pass bands for various filters
- Explain how a band-reject filter functions
- Design very simple filter circuits

FILTER OPERATION AND CLASSIFICATION

12.1 Filter Operation

Figure 12–1 shows the basic concept of filtering. Note that with different frequency components coming into the filter, only one frequency component is allowed to leave. The low-pass filter blocks the high frequency (100 kHz) and passes the low frequency (1 kHz). The high-pass filter works in an opposite manner in that it passes the high frequency and blocks the low frequency.

FIGURE 12–1 Filter concept indicating that a filter passes the correct frequency to a load.

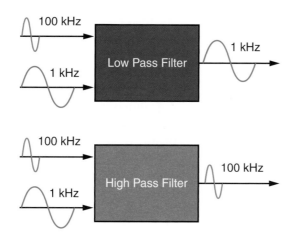

12.2 Filter Classification

There are many types of electrical filters used for different applications:
- Direct current combined with alternating current (not an actual filter—DC is superimposed with the AC)
- Transformer coupling (to block DC)
- Capacitive coupling (to block DC)
- Low-pass filters (blocks higher frequencies)
- High-pass filters (blocks lower frequencies)
- Band-pass filters (passes only a consecutive area of frequencies)
- Band-reject filters (blocks only a consecutive area of frequencies)
- Other circuit filters: Interference and Resonant

Transformer coupling of amplifier stages is often used in audio amplifiers. The concept is that transformers do not operate on DC but do respond to the fluctuating voltage of an AC signal. We know that transformers must have motion of the magnetic field created by the AC waveforms. The transformer does not transform a steady DC level. Therefore, the DC level of one stage is blocked, or ignored, by the transformer and the AC level is transformed to a higher, or possibly lower, level for the next stage or to an output device. See Figure 12–2.

FIGURE 12-2 Transformer coupling of amplifier stages.

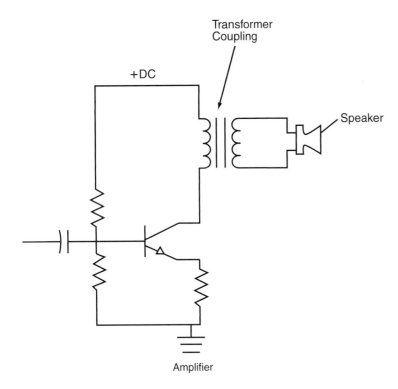

Transformer Coupling

+DC

Speaker

Amplifier

Capacitive coupling or RC filtering is also used when the DC must be "filtered out" or blocked and the AC part of the amplified voltage is allowed to pass. The filter or coupling capacitor is an inexpensive method to block the DC level. See Figure 12–3 for the placement of a DC coupling capacitor. The concept used is based on the idea that a capacitor blocks DC and passes AC. As you saw in earlier chapters, a capacitor can charge to a DC value and then appears as a counter voltage to the applied voltage. If the applied voltage does not change, the capacitor voltage does not change and appears as an open because no current flows. This feature allows the capacitor to appear as an open to a DC level voltage, but an AC voltage rising above, and falling below, the steady DC level appears as an AC that is oscillating above and below a steady reference point. Therefore, the AC component appears to pass through the capacitor and the DC value is blocked by the capacitor. The coupling of the stages of an amplifier passes the AC signal to the next stage.

FIGURE 12-3 Capacitive coupling or RC filtering of amplifier stages.

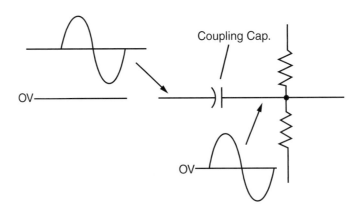

Coupling Cap.

0V

0V

LOW-PASS FILTERS

A simple low-pass filter uses an inductor in series with a load resistor. A low-pass filter is shown in Figure 12–4. At low frequencies, the chosen coil's inductive reactance will be low and very little voltage will be dropped across the inductive reactance. Most of the source voltage will be dropped across the load (R_L).

FIGURE 12-4   Schematic of a low pass RL filter and graph representation of the output.

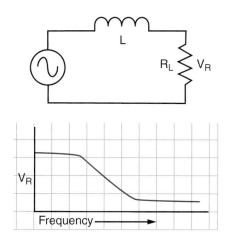

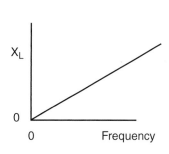

Remember that $X_L = 2\pi fL$, and with an impedance triangle with very little reactance, almost all of the voltage drop will be due to resistance.

As the frequency increases, the inductance does not change but the X_L increases. The result of the increase in X_L is that more voltage is dropped across the inductor and less voltage is dropped across the load. As you can see in the accompanying graph in Figure 12–4, the voltage across the load (R_L) starts to decrease as the frequency increases and the X_L increases, until very little of the source voltage is dropped at the load. A common application for choosing chokes (coils) is if the X_L value (measured in ohms) is 10 times greater than the load resistance (also measured in ohms), there is virtually no signal voltage across the load.

The opposite is also true. If the value of X_L is 10 times less than the value of the resistance, most of the signal will reach the load as is the case at the low starting frequency. Typically, low-pass filters are designed to deliver most of the signal to the load at lower frequencies. Use Figure 12–4 as the schematic using values of 1000 Ω for the resistor and .16 Hz for the coil. The circuit has 100 V at 100 Hz applied.

$$X_L = 2\pi fL$$

$$X_L = 2 \times 3.14 \times 100 \times .16$$

$$X_L = 100.48\ \Omega\ @\ 100\ \text{Hz}$$

Using a voltage vector for a series circuit, the voltage drop across the 1 k Ω resistor will be 99.5 V and the voltage drop across the 100 Ω X_L will be 9.9 V as shown in Figure 12–5.

When the frequency increases to 2000 Hz, the shape of the impedance and the voltage vector changes to that shown in Figure 12–4, and explained in the subsequent example.

$$X_L = 2\pi fL$$

$$X_L = 2 \times 3.14 \times 2000 \times .16$$

$$X_L = 2010\ \Omega\ @\ 2000\ \text{Hz}$$

FIGURE 12-5

Voltage vector representing values in Figure 12–4 @ 100 Hz.

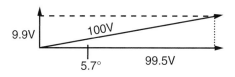

FIGURE 12–6

Voltage vector representing voltage values for Figure 12–4 @ 2000 Hz.

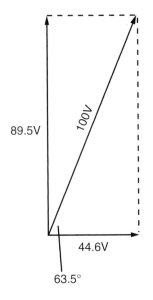

The voltage vector now appears as in Figure 12–6, with an angle of lag of 63.5°. This causes a voltage drop of 89.5 V across the coil and 44.6 V across the resistor, trapping a larger amount of voltage at the choke coil (choking off the voltage supply) and allowing a smaller amount of voltage to be passed to the load resistor at higher frequencies.

Now increase the frequency to 10 kHz (10,000 Hz) to view the impedance and voltage triangles as seen in Figure 12–7. As you can verify by the mathematics below, a large portion of the source voltage appears across the "choke" and a very small value appears across the load resistor.

FIGURE 12–7 Impedance triangles and the voltage vector representing values for Figure 12–4 @ 10,000 Hz.

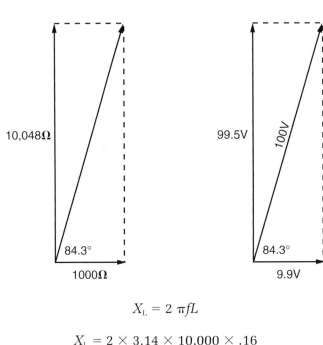

$$X_L = 2\,\pi f L$$

$$X_L = 2 \times 3.14 \times 10{,}000 \times .16$$

$$X_L = 10{,}048\ \Omega \ @ \ 10{,}000\ \text{Hz}$$

12.3 Half-Power Point

Half-power point
The point at which the load resistance is dissipating half the maximum power that could occur at the load.

To find the **half-power point** of a filter we do a quick calculation regarding X_L and R_L. From the previous example, it is easy to ascertain that half-power is defined as the point at which the load resistance is dissipating half the maximum value. It is easy to assume this will be true when X_L equals the value of the resistance, and, in fact, this is the case.

To prove it, start by considering the fact that the *maximum* power will be dissipated by the resistor when the voltage on the load resistor is equal to the supply voltage. This is true because at this point the voltage across the resistor is maximum, and therefore the current flow through the resistor is greatest, and the power dissipation is maximum. This maximum power will occur at the maximum voltage drop across the load resistor, which is represented as E_{sup}.

Now the formula for the maximum power is given by:

$$P_{max} = \frac{E_{sup}^2}{R_L}$$

Solving for E_{sup} gives:

$$E_{sup}^2 = P_{max} \times R_L$$

and:

$$E_{sup} = \sqrt{P_{max} \times R_L}$$

The voltage for one-half power is given by:

$$E_{1/2} = \sqrt{\left(\frac{P_{max}}{2}\right) \times R_L}$$

Substituting for P_{max} yields:

$$E_{1/2} = \sqrt{\frac{\left(\frac{E_{sup}^2}{R_L}\right)}{2} \times R_L} = \sqrt{\frac{E_{sup}^2}{2}}$$

The voltage drop across the resistor can also be calculated using the voltage divider formula:

$$E_{1/2} = \frac{R_L}{\sqrt{X_L^2 + R_L^2}} \times E_{sup}$$

This means that:

$$\sqrt{\frac{E_{sup}^2}{2}} = \frac{R_L}{\sqrt{X_L^2 + R_L^2}} \times E_{sup}$$

Squaring both sides yields:

$$\sqrt{\frac{E_{sup}^2}{2}} = \frac{R_L^2}{\sqrt{X_L^2 + R_L^2}} \times E_{sup}^2$$

Dividing both sides by E_{sup}^2 gives:

$$\frac{1}{2} = \frac{R_L^2}{X_L^2 + R_L^2}$$

Cross-multiplying gives:

$$X_L^2 + R_L^2 = 2R_L^2$$

And subtracting R_L^2 from both sides yields:

$$X_L^2 = R_L^2$$

Notice that this will be true even if X_L is negative. Of course, inductive reactance cannot be negative; however, capacitive reactance can be, so the same formula holds for a series RC circuit.

At half power, the voltage across the load is the same as the voltage drop across the coil as can be seen in the impedance triangle and voltage vector. The angle of lag is 45° and the voltage across the load resistor is .707 times the line voltage. The current flowing through the load is .707 times the maximum possible value. The product of 70.7% of the voltage at 70.7% of the current is approximately 50% of the power.

TechTip!

When a circuit allows one-half of the power to pass through a filter and appear at the load, the filter is said to be at the $\frac{1}{2}$ power point. The object of most filters is to allow $\frac{1}{2}$ or more of the power to appear at the load and be able to block or stop $\frac{1}{2}$ of the circuit power at the filter, based on the designed frequency. The desired frequency should allow $\frac{1}{2}$ or more of the power to pass and be able to block at least $\frac{1}{2}$ of the undesired power. This is typically the point where 70.7% of the source voltage appears at the load. If 70.7% of the voltage appears at the load, it is assumed that 70.7% of the current appears at the load. 70.7% voltage × 70.7% current = 50% of the volt amps or power, or $\frac{1}{2}$ power point.

All frequencies that lose less than the one-half maximum power at the coil are considered to pass to the load. All frequencies that produce more than one-half power dissipation at the coil are rejected and never reach the load.

12.4 Inductive Low-Pass Filter

As an example, design a low-pass filter that will pass all frequencies up to 400 Hz to a load of 600 Ω. The half-power point occurs when $X_L = R_{Load}$. With these two values equal, one-half of the power or more is delivered to the load at any frequency up to 400 Hz. To find the henry value of the inductor when X_L equals 600 Ω at 400 Hz, use the following formula:

@ half-power point at 400 Hz: $X_L = R_L$, $X_L = 600\ \Omega$

$$X_L = 2\ \pi f L$$

or

$$L = \frac{X_L}{2\ \pi f}$$

$$L = \frac{600\ \Omega}{6.28 \times 400\ \text{Hz}}$$

$$L = .239\ \text{H}$$

12.5 Capacitive Low-Pass Filter

Another low-pass filter is shown in Figure 12–8. This filter has a capacitor in parallel with the load resistor. The circuit has a series resistor R_S in order to drop the voltage that is not delivered to the load. At low frequencies, the capacitor will exhibit a high reactance, and most of the circuit current will flow through the load resistor rather than the capacitor branch. With R_L larger than R_S, a greater percentage of the line voltage will drop across R_L.

FIGURE 12–8 Low pass filter with capacitor across load.

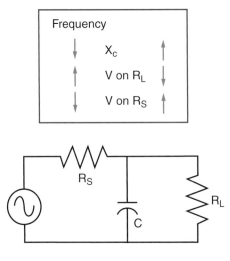

If the frequency increases, the capacitor's reactance decreases and more current flows through the capacitor. With this condition the series resistor will drop more voltage and cause the load voltage to decrease. This is depicted in the chart associated with Figure 12–8.

As the frequency is increased, the capacitive branch will draw greater current because of the decreasing value of X_C. The circuit impedance will decrease as the capacitive current increases. The circuit current will increase and as a result, greater current will flow through the series resistor (R_S). This causes the voltage across R_S to increase, with the voltage across the parallel circuit of the filter capacitor and R_L to decrease. See Figure 12–9 for a graphical representation of this filter's effect on load voltage. This type of filtering is often used across the power line to reduce high-frequency interference. It will filter out all of the higher frequencies and allow normal 60 Hz frequency power to pass.

FIGURE 12–9 Graph of load voltage versus frequency for the circuit in Figure 12–8.

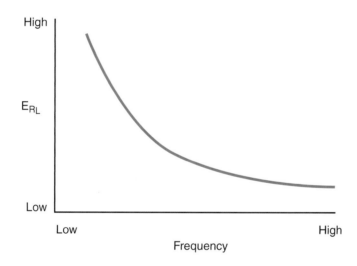

12.6 Combined LC Low-Pass Filter

Another example of a low-pass filter is shown in Figure 12–10. This filter consists of an inductor in series with a capacitor and load resistor in parallel as the load. As you surmise, at low frequencies the X_L is low and the X_C is high by comparison. At the lower frequencies, very little voltage will be dropped across the inductor. Most of the source voltage will be across the very high reactance of the capacitor and the load. As the frequency is increased, more of the source voltage will be across the inductor, and the voltage across the capacitor and the load will decrease. This filter has a sharper slope and is more efficient than the filter in Figure 12–8 because both the inductor and the capacitor serve to reduce the voltage across the load simultaneously as the frequency increases.

When the circuit reaches the resonant frequency ($X_L = X_C$), the output signal will peak. This will be followed by a sharp drop in the output signal (voltage) as the frequency continues to rise. A graph of this load voltage vs. frequency is shown in Figure 12–10.

TechTip!

Some intercom systems use the power line to carry the signal, and this filter would block these low frequencies of 60 Hz. Other control systems use the power line to carry the signals for energy control, lighting control, etc. The filter is set at the receiver to only pass the desired signal frequency and block the normal 60 Hz line frequency. By using a specific capacitor in series, a simple and cheap filter is achieved.

FIGURE 12–10 Diagram of low pass filter that utilizes an LC circuit for sharper cutoff, and graph of output voltage versus frequency for the circuit shown in the diagram.

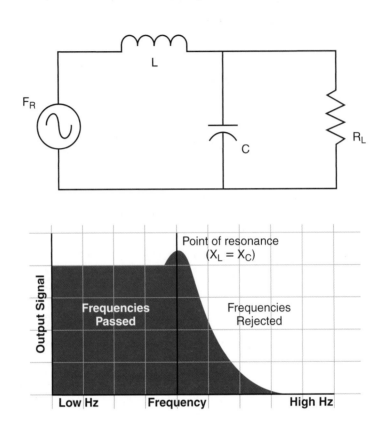

HIGH-PASS FILTERS

A typical high-pass filter circuit schematic is shown in Figure 12–11. A simple high-pass filter consists of a capacitor in series with the load. This configuration is easily recognized as a simple series RC circuit.

FIGURE 12–11 Schematic of simple RC high pass filter.

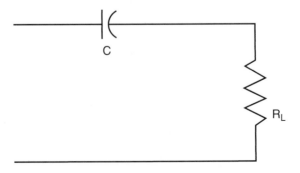

At low frequencies, the X_C is very high, and the capacitor will drop the most voltage. As indicated in the impedance and voltage triangle of Figure 12–12, as the frequency is increased, less voltage will be dropped across the capacitor and more voltage will be dropped across the load. This type of filter can be used to eliminate interference (at power line frequencies) from entering the load. In order to design the filter to trap a specific frequency, we will need to find the components that will produce the ½ power point for the circuit. In the first example we will trap any frequencies that are 1000 Hz and below. See the graph of voltage vs. frequency for this type of filter depicted in Figure 12–11.

FIGURE 12–12 Diagram of voltage and impedance triangles, and graph of voltage versus frequency for the circuit shown in the diagram.

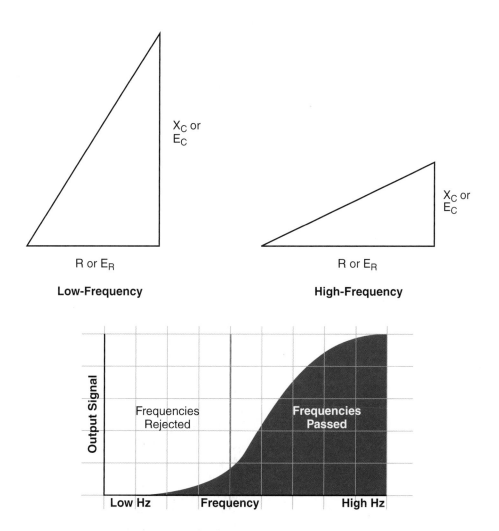

As you saw previously, the half-power point for this filter will occur when the capacitance reactance (X_C) equals the load resistance at the desired low cutoff of the filter frequency. First find the capacitor that will create the desired reactance to produce one-half power at 1000 Hz.

Example

At 1 kHz and a load of 1 kΩ

If: $X_C = R_L$

Then: $X_C = 1000\ \Omega$ @ 1000 Hz

Therefore: $X_C = \dfrac{1}{2\ \pi f C}$ where C is in farads.

Solving for C:

$$C = \frac{1}{2\ \pi f X_C}$$

$$C = \frac{1}{6.28 \times 1000 \times 1000}$$

$$C = \frac{1}{6{,}280{,}000}$$

$$C = .159\ \mu F$$

Then find the reactance at 60 Hz and the same microfarad capacitor:

$$X_C = \frac{1}{2\ \pi f C}\quad \text{where } C \text{ is in farads}$$

$$X_C = \frac{1}{2\ \pi \times 60 \times .000000159}$$

$$X_C = \frac{1}{6.28 \times 60 \times .000000159}$$

$$X_C = 16{,}691.4\ \Omega$$

As previously shown, the voltage triangle for the series RC circuit is very different at 1000 Hz than it is at 60 Hz. The opposition offered by the capacitor is over 16 times the opposition of the load at 60 Hz and most of the voltage is trapped at the capacitor and does not appear across the load.

12.7 Capacitive High-Pass Filter

When the X_C value is one-tenth as great as the load resistance, there is very little, or essentially no, voltage signal across the capacitor. As an example, consider a filter that will pass all frequencies above 400 Hz to a 600 Ω load. The first step is to calculate the value of C. Again, we assume a half-power point where the load opposition equals the series drop component opposition. In other words, X_C equals R load at 600 Ω. At this point at 400 Hz, one-half of the power will be delivered to the load. At any higher frequencies more of the circuit power is delivered to the load.

The capacitance value equals

$$C = \frac{1}{2\,\pi f\,X_C} \qquad C = \frac{1}{6.28 \times 400 \times 600} \qquad C = \frac{1}{1,507,200} \qquad C = 0.66\ \mu F$$

As frequencies increase above this value, X_C decreases even further and less voltage is dropped at the capacitor and more at the load resistor in series.

12.8 Inductive High-Pass Filter

In Figure 12–13 an inductor is in parallel with the load resistor. At low frequencies the inductive reactance (X_L) is low, allowing most of the circuit current to flow through the inductor. Using $E_L = I \times X_L$ means that the parallel voltage drop is low and the load voltage is low. To exaggerate this effect, imagine the frequency is 0 Hz. The X_L of the coil is 0 Ω, or in other words a short circuit with no voltage drop. As the frequency increases and the inductive reactance (X_L) increases comparatively more current flows through the parallel resistive load. Since the resistance did not change, the voltage across the load increases. When the value of the inductive reactance (X_L) becomes 10 times the value of the load resistor, essentially all the signal current flow is through the load resistor and the inductor no longer has any effect on the circuit. To take this to an extreme, assume the frequency goes to infinite hertz. The X_L becomes so great that it appears as an open circuit. See Figure 12–14 for a sample of the current triangles of the parallel R_L circuit. See Figure 12–13 to see the effects of increasing frequency on the voltage across R_L.

FIGURE 12–13 Schematic for inductive high-pass filter using RL and graph showing voltage versus frequency curves for the schematic.

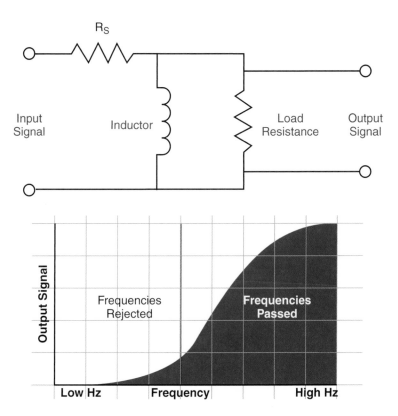

FIGURE 12-14 Current vector representation for the circuit in Figure 12–13.

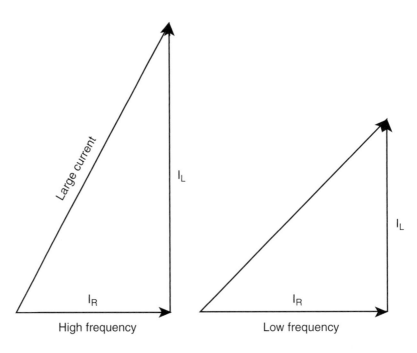

12.9 Combined LC High-Pass Filter

In Figure 12–15 the series resistor (R_S) is replaced with a capacitor. Again, at the low frequencies more of the source voltage will be across the high X_C of the capacitor (because of its high reactance at low frequency) and less voltage will be across the inductor and the load (because of the low reactance of the inductor at low frequency).

FIGURE 12-15

Schematic of simple high-pass filter using LC components and graph showing voltage versus frequency curves for schematic.

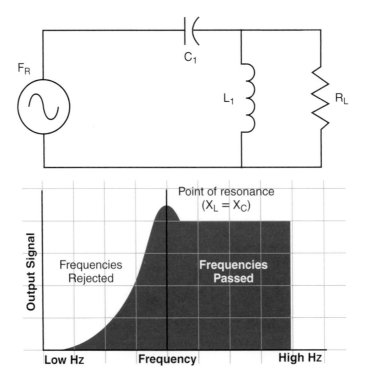

As frequency is increased, the capacitive reactance decreases, the inductive reactance increases, and a greater proportion of the source voltage drops across the parallel coil and resistor combination. Therefore, the voltage across the load increases with the increasing frequency. There is a sharper cutoff of frequencies because of the combined effect of falling capacitive reactance and simultaneously rising inductive reactance (Figure 12–15). Remember, you can determine the half-power point where the frequency passes enough power to the load at the low frequency cutoff point by calculating the value of frequency when X_L equals X_C. The value of resonant frequency (F_R) is when X_L and X_C are equal or:

$$F_R = \frac{1}{2 \pi \sqrt{LC}}$$

BAND-PASS FILTERS

Bandwidth refers to the resonant frequency and the frequencies on either side of the resonant frequency that allow the voltage to be delivered to the load rather than being trapped out, or blocked. Specifically, bandwidth refers to the usable range of frequencies that allow a signal to pass from input to the filter to the output of a filter circuit, which is designated as the load. The span of frequencies results in the range, or band, of frequencies. The bandwidth of the band pass filter is considered the range of frequencies that allows output voltage at the load to be at least 70.7% of the maximum output voltage that occurs at exact resonance. As previously explained, this is considered to be the ½ power point. A series resonant bandpass filter would appear as in Figure 12–16 with accompanying graph. This tells us that the lower end of the frequencies that are allowed to pass will provide 70.7% of the maximum voltage.

FIGURE 12–16

Schematic for bandpass circuit using high Q components and graph showing voltage versus frequency for the schematic.

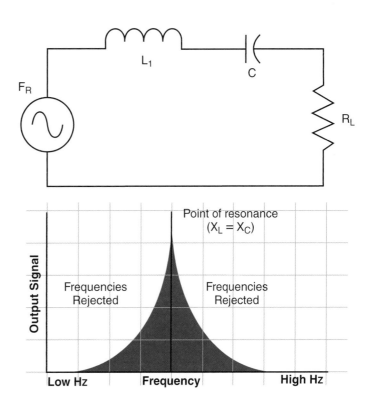

The frequency at the upper end will also allow 70.7% of the maximum voltage to pass. These frequencies are referred to as the lower and upper cutoff frequencies respectively. The exact middle of the curve is the resonant frequency, which allows maximum voltage to appear at the load. Therefore, this is the band of frequencies that will pass at least 70.7% of the source voltage to the load. The formula for bandwidth is: Bandwidth = Upper cutoff frequency − lower cutoff frequency.

12.10 Series RLC Bandpass Filter

A series RLC bandpass filter can be used to create a filter network. The concept of series circuit resonance is used for a series LC circuit as in Figure 12–16. With this type of series circuit, the resonant frequency will cause the X_L and X_C of the coil and capacitor to cancel each other to create a minimum circuit impedance and, therefore, allow a large circuit current to flow. This large circuit current allows a large voltage drop across the RL at resonant frequency, and permits a large percentage of the input voltage, and power, to be delivered to the load. All other frequencies will drop voltage at the series components of inductance or capacitance and do not reach the load. This situation is depicted in the graph associated with Figure 12–16.

You can see by the graph in Figure 12–16 that at low frequencies there is a large volt drop at the series capacitor. This occurs because of the large capacitive reactance in comparison to the inductive reactive component. At resonance, the two reactive components cancel and the Z of the circuit is low. This causes a large current to flow through the load and a large voltage is dropped at the load. As the frequency increases, the X_C decreases and the X_L increases. As X_L increases, the Z of the series circuit increases, the circuit current decreases and less current flows through R_L, producing a lower load voltage and current. The resonant frequency (F_r or F_0) for this circuit is determined by:

$$f_r = \frac{1}{2\pi\sqrt{LC}}$$

The bandwidth of a filter is affected by the Q factor of the circuit. The Q factor is the ratio of the reactive power in the inductor to the true power of the full circuit resistance. The Q of the coil is determined by the $\frac{\text{VAR}}{\text{watts}}$ of the circuit at resonance. This can also be expressed as $\frac{\text{XL}}{\text{R}}$. A higher value of Q results in a smaller bandwidth. If the Q is higher, the XL is large compared to the resistance of the entire circuit. As the frequency goes above resonance, the voltage will drop on at the series coil due to XL. Likewise, if the frequency drops below resonance, the effect of the capacitor is more apparent. The ratio of change is higher with a larger Q of the circuit, which creates a smaller bandwidth. The formula for Bandwidth (BW) = Frequency at resonance $\frac{\text{(fr)}}{\text{Quality (Q)}}$.

To find the actual lower cutoff frequency, use: $F_1 = \dfrac{F_R - \text{BW}}{2}$ and for the upper cutoff frequency: $F_2 = \dfrac{F_R + \text{BW}}{2}$.

See the Example that follows for Figure 12–16.

To determine the bandwidth (*BW*), the Q of the circuit must first be determined:

$$Q = \frac{X_L}{R}$$

$$BW = \frac{f_R}{Q}$$

$$F_1 = f_R - \frac{BW}{2}$$

$$F_2 = f_R + \frac{BW}{2}$$

Example

X_L is 1000 Ω, resistor is 100 Ω, X_C = 1000 Ω at resonant frequency for this example is 500 Hz:

$$Q = \frac{1000}{100}$$

$$Q = 10$$

$$BW = \frac{500 \text{ Hz}}{10}$$

$$BW = 50 \text{ Hz}$$

$$F_1 = 500 - \left(\frac{50}{2}\right) \text{ or } 475 \text{ Hz for lower limit}$$

$$F_2 = 500 + \left(\frac{50}{2}\right) \text{ or } 525 \text{ Hz for upper limit}$$

Note that *f*1 (minimum frequency) and *f*2 (maximum frequency) are the minimum and maximum frequencies for these components.

A general rule of thumb relative to the Q value of the components is that a circuit with high-Q components equals a small, or narrow, bandwidth. A circuit with low-Q components equals a wide bandwidth.

12.11 Band-Pass Filter with Parallel Load

Parallel LC band pass filters use a filter network as in Figure 12–17. The primary use for this type of circuit is to pass a specific frequency, so that the load receives a high percentage of the source voltage and power. As learned in previous chapters, a tank circuit exists when a circuit contains a coil and a capacitor in parallel. In the tank circuit the inductive current and the capacitive current are 180° out of phase with the same voltage across each component. When the LC circuit is at resonance the currents cancel each other out and the circuit current from the source goes to a minimum, creating a maximum circuit impedance. This maximum value of circuit impedance occurs only at resonant frequency for the coil and capacitor in the circuit. At this resonant frequency, a small amount of current flows through R_S, so only a small voltage drops on R_S and most of the desired voltage drops across the load, as depicted in the associated graph in Figure 12–17.

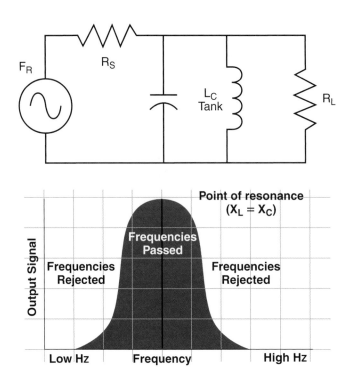

FIGURE 12–17 Schematic for a bandpass filter using low Q components and graph showing voltage versus frequency for the schematic.

In Figure 12–18, the components of the LC parallel filter are considered low Q. In a parallel RLC circuit the Q of the circuit is expressed as the ratio of the R to the XL of the circuit. If the R of the load (RL) is low, the Q of the filter is low. The bandwidth is still expressed as $\dfrac{FR}{Q}$, so the bandwidth is comparatively larger with low resistance loads. As before, a higher value of Q results in a smaller bandwidth. This means that with a high current draw of a load, the Q decreases and the frequency bandwidth becomes greater.

FIGURE 12–18 Band pass filter with high resistance R_L.

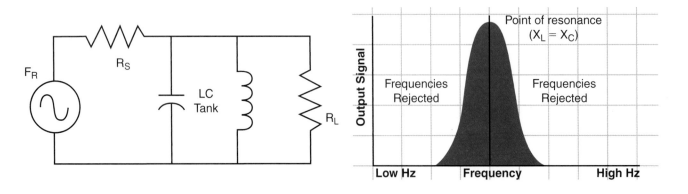

In Figure 12–18 the load resistance is high, creating a small current load, and the Q of the filter circuit is higher. Again, the bandwidth of the circuit is $\frac{FR}{Q}$, meaning the bandwidth of the filter is smaller, as depicted in the graph of Figure 12–18.

BAND-REJECT FILTERS

A band-reject filter, or band-stop filter, as illustrated in Figure 12–19, is an example of a band-reject filter with high-Q components. This circuit consists of an inductor and capacitor connected in parallel with each other and in series with the load resistor. (Notice that to go from band-pass to band-reject filtering, the load was moved from the parallel to the series position with the LC tank portion of the circuit.) At resonance, the impedance of the resonant circuit will be at maximum, and a minimum current will flow to the load. The high impedance is due to the inductive and capacitive currents canceling each other. At resonance, very little current flows to the load and therefore very little of the source voltage is dropped across the load resistor. At frequencies below or above resonance the LC parallel trap allows more line current to flow. As a result of this circuit response, more current flows through RL and more of the source voltage is delivered to the load. In other words, any frequency other than resonance will deliver voltage to the load.

TechTip!
....

Filters are used throughout the electrical industry to reduce or eliminate unwanted frequencies or pass the desired frequencies. They are used extensively in electronic equipment to filter DC ripple from power supply outputs or audio circuit tone and sound controls. By making some of the capacitive or inductive components variable, the resonant frequency points can be adjusted to let through specific broadcast frequency signals for radio and TV stations.

FIGURE 12–19 Schematic for a band-reject filter with high Q components and graph for band-reject filter using high Q compoenents as shown in the schematic.

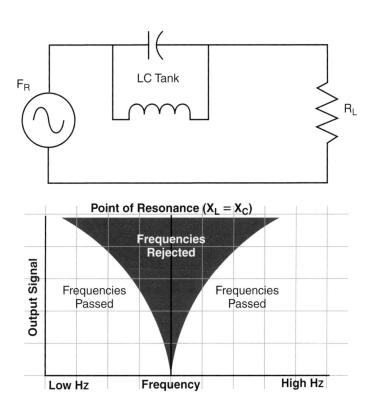

12.12 Band-Reject Filter with Parallel Load

Figure 12–20 is an example of a band-reject filter circuit with low-Q components. In this case, the load is connected in parallel to the resonant series LC circuit. A resistor (R_S) is connected to this configuration in series with the source. At resonance, the inductive reactance and the capacitive reactance cancel each other because their reactances are 180° out of phase. This places the load resistor (R_L) in parallel with a very low impedance, and the current through the LC filter is high. The current flow through the series resistor (R_S) will be at maximum at this point, causing a large volt drop across the R_S, thereby producing a minimum voltage across the load resistor. Figure 12–20 shows that the bandwidth rejection increases for low-Q components. As the frequency is any value other than resonant, the LC series circuit Z increases and the line current decreases. As the line current decreases, less voltage is dropped on R_S and more on R_L, thereby passing the frequencies other than resonant.

FIGURE 12–20 Schematic of band-reject filter using low Q components and graph of voltage versus frequency for the schematic.

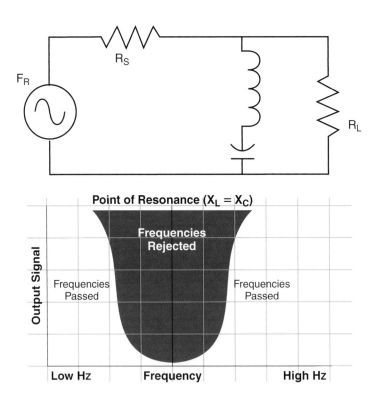

COMPLEX FILTERS

12.13 T-Type Filter

More often than not, filters are more complex than the ones described in the previous examples. Figures 12–21 and 12–22 show two T-type filters (resembling the capital letter T in schematic form).

FIGURE 12–21 Schematic of T-type filter used to pass high frequencies and graph of voltage versus frequency for the schematic.

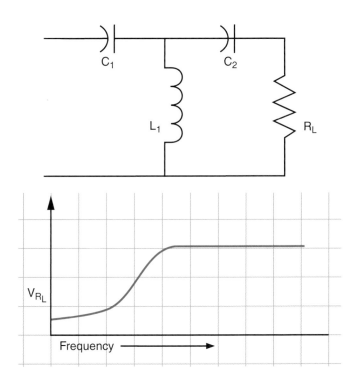

FIGURE 12–22 Schematic of low pass T-type filter and graph of voltage versus frequency for the schematic.

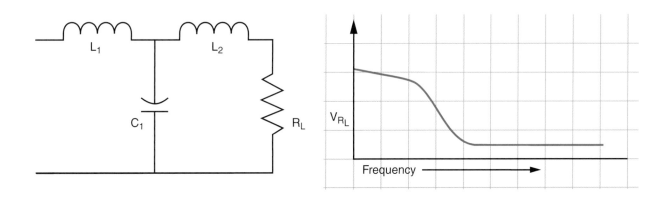

The filter in Figure 12–21 is a high-pass filter, and the one in Figure 12–22 is a low-pass filter. Several sections of T-type filters may be joined in some applications. These types of filters improve filtering by increasing sharpness of the cutoff frequencies.

The high frequencies are allowed to develop across the load by the inductor (L_1) (Figure 12–21). At low frequencies, the inductor essentially acts as a shunt across the load. See Figure 12–21 to see the effects of a filter that passes only the higher frequencies to the load. The concept is one of trapping the low frequencies at the capacitors that have high reactance at low frequencies with a larger volt drop across the coil with higher reactance at high frequencies.

In Figure 12–22, the low frequencies are allowed to pass by the inductors (L_1 and L_2), and the capacitor acts as a high impedance for the load. Figure 12–22 shows the effects of the filter passing the low frequencies and trapping the higher frequencies.

12.14 π-Type Filter

Figure 12–23 is an illustration of a π (pi)-type filter. This filter is commonly found in power supplies to smooth the rectified AC wave to smoother DC. The representation in Figure 12–23 of an input to the filter shown is the 120 Hz pulses of a full-wave rectifier supplied by a 60 Hz AC source and full-wave rectification. The capacitors usually have very large capacitance (10 µF to 1000 µF), and at 120 Hz also have low reactance. The low reactance means that the current to the capacitors is comparatively large. The inductance, at the same time, has a value in the range of 10 to 100 H. This combination pre-sents a high impedance (opposition) to the changing voltage and current of the rectified input. The capacitors charge to the peak value of the rectified voltage.

FIGURE 12–23 π-Type low pass filter schematic and graph displaying voltage input and output.

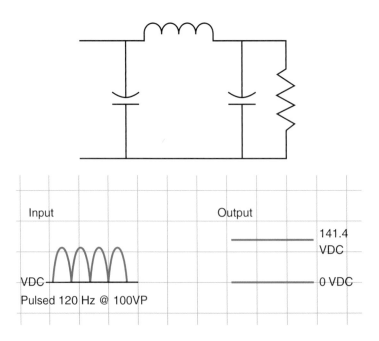

If the input is 100 V_{RMS} (Voltage (Root Mean Square)), the output will be 141.4 VDC (Voltage (Direct Current)). This filter would be classified as a low-pass filter, allowing low frequencies (such as DC with no frequency) and blocking the 120 Hz ripple frequency.

A high-pass π-type filter is shown in Figure 12–24. In this circuit, the inductors now form the legs of the π, with the capacitor between them. Low frequencies will be blocked by the capacitor and shunted through the inductors around the load. The high-frequency components of the input will develop very little voltage across the series capacitor (C_1), allowing most of the voltage to be produced across the load (R_L) at high frequency.

FIGURE 12–24 High frequency pass π-type filter.

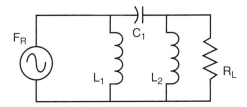

12.15 Other Complex Filters

Even more complex filters are shown in Figures 12–25 and 12–26. Figure 12–25 is a complex band-pass filter, and Figure 12–26 is a complex band-reject filter.

Notice that the band-pass filter has the resonant circuit components in series with the load (R_L) and that the band-reject filter has the tank circuit in series with the load (R_L). These filters can be tuned to pass a given band of frequencies and reject interfering frequencies operating very close to the desired band. These combinations may also be combined to form even more complex filters. Engineers specializing in this field are usually the designers of these types of filters.

FIGURE 12–25

Complex frequency filtering circuit using a combination of filter to provide band pass filtering.

FIGURE 12–26

Complex filtering circuit used to provide band-reject filtering.

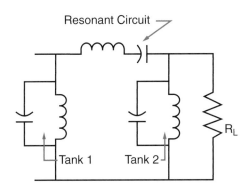

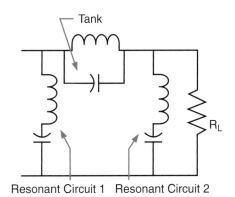

SUMMARY

In general, characteristics for high-pass filters are the following:

- Capacitance is connected in series with the load. The X_C will be low for high frequencies to be passed to the load, while low frequencies are blocked by a high X_C.

- Inductors are connected in parallel with the load. The shunt X_L will have a high reactance to high frequencies, which will cause current to flow through R_L, while low frequencies are bypassed through the inductor.

Characteristics for low-pass filters are the following:

- Inductors are in series with the load. The high X_L for high frequencies serves as a choke or block, while low frequencies are passed to the load.

- A bypass capacitor is connected in parallel with the load. The high frequencies are bypassed by a small X_C, while low frequencies are not affected by the shunt path and flow through R_L.

Filters are becoming increasingly important in electrical work. Computers radiating signals or information that escape through their power source allow for anyone with access to a receptacle or breaker box to collect sensitive information. In worst case scenarios, such "information leaks" can jeopardize government and corporate security.

REVIEW QUESTIONS

1. Discuss the basic objectives of low-pass, high-pass, band-pass, and band-reject filters.

2. What are the underlying concepts that make it possible to create filters using R, L, and C components?

3. How is the half-power point related to the voltage drop across the load resistor in a high-pass or a band-pass filter?

4. Discuss the applications for filters using low-Q components; discuss those for high-Q components.

5. Explain bandwidth in terms of Q and list the upper and lower limit formulas in a series RLC bandpass filter. Give a numeric example of Q, BW, f_1, and f_2.

6. Explain the two types of π filters and their uses.

7. What is meant by a band-stop filter circuit?

8. Draw a schematic diagram of a typical simple low-pass filter.

PRACTICE PROBLEMS

1. Name five types of filters for which you can recognize and discuss the output frequency curve.

Assume a low-pass RL filter for Problems 2–4:

2. The load resistor will have a voltage across it when the frequency is ____ the half-power point.

3. The half-power point is defined as the point where X_L is equal to ____.

4. What is the cutoff frequency for a low-pass R–L filter having a 3.3 kΩ resistor and a 1 mH coil?

Assume a low-pass RC filter for Problems 5 and 6:

5. Low-pass RC filters have the capacitor connected in ____ with the load resistor.

6. Design a 2 kHz low-pass RC filter with a 1 kΩ series resistor; that is, find C.

Assume a low-pass LC filter for Problems 7–9:

7. In a low-pass LC filter, the output signal will peak when $X_L =$ ____ (resonant frequency).

8. What is the peak frequency for the speaker in Figure 12–27?

FIGURE 12–27 Figure to be used to answer Problems 8 and 9.

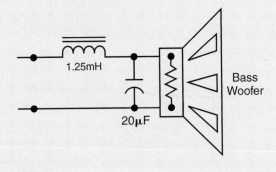

1.25mH

20μF

Bass
Woofer

9. Find the new resonant frequency if the capacitor were changed to a 10 μF capacitor.

Use a high-pass RC filter for Problem 10.

10. Using a .1 μF capacitor, find the resistance needed to make a high-pass 300 Hz filter.

Use a high-pass RL filter for Problem 11:

11. Find the cutoff frequency for a high-pass R_L filter with a 22 k Ω resistor and a 500 μH coil.

Use a high-pass LC filter for Problem 12:

12. The output of a high-pass LC filter increases as the X_L ____.

Use a band-pass filter for Problems 13 and 14:

13. What is the center frequency for a series band-pass filter with a 100 μH coil and an 800 pF capacitor?

14. Find the center frequency for a parallel band-pass filter with a 100 μH coil and a 100 pF capacitor.

Use a band-reject filter for Problems 15–17:

15. Find the center frequency for a series band-reject filter having a 42.9 mH coil and a .0041 μF capacitance.

16. If the Q of the circuit in the previous problem is 3, what is the bandwidth?

17. Find f1 and f2 for the previous circuit.

13

Generators

OVERVIEW

English physicist Michael Faraday discovered the basic theory of electrical generation in 1831. He discovered that a magnetic field can be used to produce an electrical current. Practically all commercial electrical power is produced using this principle.

Mechanical energy delivered from raw manpower, water, wind, fossil fuels (coal and oil), and nuclear reactors is used to turn the rotor of a generator. Some of these energy sources, such as fossil fuels and nuclear reactors, must first be converted into another form such as steam, in order to turn the rotor of a generator. This chapter will acquaint you with both DC and AC generators.

Although AC is available at most sites, DC has applications that AC cannot support. DC has been used for precise speed control of motors and for large DC motors found in many industrial and manufacturing applications. DC motors and generators are also used in locomotives. Where DC is required, electricians will often find DC generators.

As variable frequency drives become more and more common, the use of DC generation and DC motors will become less frequent. However, the understanding of DC equipment is still necessary to a professional electrician. Eventually, DC power distribution will be found only in those places where extremely high torque per horsepower is required (DC motors) or where the nature of the work requires DC, such as in electrolysis applications.

This chapter covers AC generation fundamentals and shows the relationship between magnetism (which you have previously studied) and the production of electrical power. It identifies the major parts of a DC or AC generator and explains the fundamental operating principles of each. Remember that DC machines will be with us for many more years, and knowledge of how they operate and how to install and repair them will be a valuable skill. AC generators are used everywhere from small portable generators, to truck mounted diesel engine driven generators, to permanent standby or emergency generators, to the very large commercial power generating stations. A thorough understanding of AC generators is needed when working in today's technical workforce.

OBJECTIVES

After completing this chapter, you should be able to:

* Identify the major parts of a DC generator
* Describe the principles of operation of the DC generator
* List the advantages of different generator connections
* Explain the parallel connections for DC generators
* Describe the operation of AC generators
* Identify key parts of the AC generators and their functions
* Mathematically determine the relationship between revolutions per minute, frequency, and the number of poles
* Explain the voltage, frequency, and power controls on an AC generator
* List the requirements for paralleling AC generators

INDUCTION BY MAGNETISM

As you studied briefly in Chapter 1 and more in Chapter 2, a voltage is induced into a conductor when it is passed through a magnetic field. The effect is called electromagnetic induction. If the conductor is connected to a complete circuit, current will flow. Figure 13–1 shows a conductor being moved downward between the poles of the two magnets. The poles of the magnets create a field that is moving from north to south. The downward motion of the conductor induces a voltage into the conductor, the polarity of which causes the current to flow and the meter to deflect in the direction shown. The electrons move because of the interactions between the magnetic field of the permanent magnets and the electromagnetic fields of the electrons themselves. Note that without movement, there is no current flow.

FIGURE 13–1 Voltage is induced when moving a conductor through a magnetic field.

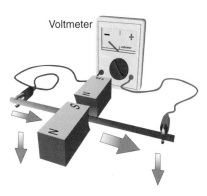

If the motion in Figure 13–1 were reversed, it would change the voltage polarity and reverse the flow of current. This is because the direction of the field interactions is opposite. The same would be true if the poles of the magnets were reversed and the motion was downward; the field interactions would be reversed, and current would flow in the opposite direction.

Remember that relative motion between the magnetic field and the conductor must exist before a voltage will be generated. This motion can be the motion of the magnetic field instead of the conductor; that is, the magnets can move instead of the conductor. Movement of the magnetic field will be an important variation in your study of AC generators.

GENERATOR CHARACTERISTICS

A generator is a device that converts mechanical energy to electrical energy. A voltage is induced in a conductor when it cuts magnetic lines of flux. This principle, called electromagnetic induction, is the underlying principle of operation for both DC and AC generators.

FieldNote!

Many people in the electrical professions learn theory based on the conventional current flow theory. When this is the case, they may refer to the "right- hand rule for generators." This, in fact, reverses the electron current flow idea and the theory of current flow from positive to negative is intact. To carry this theory further we use the right-hand rule for coils and that leads to the left-hand rule for motor direction. Both the electron flow practitioner and the conventional current flow practitioner end with the same results. Confusion ensues, however, if you use conventional flow for part of an analysis and electron flow for the other part.

Figures 13–2 through 13–5 show how this principle works. As the loop is rotated and cuts through the magnetic lines of flux, a voltage is induced in the loop. An easy way to remember the relationship between motion, electromagnetic lines of force, and current flow is to use Fleming's left-hand rule for generators, shown in Figure 13–6. This rule is used for electron flow theory explanations.

FIGURE 13–2 Position 1 with no induced voltage.

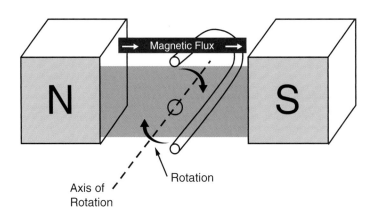

FIGURE 13–3 Position 2 with maximum induced voltage.

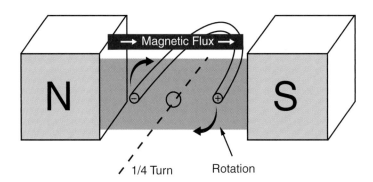

FIGURE 13–4 Position 3 with no induced voltage.

FIGURE 13–5 Position D with maximum induced voltage.

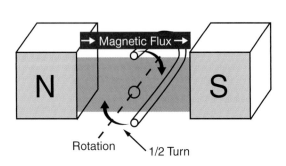

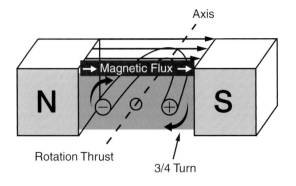

FIGURE 13-6 Fleming's left hand rule for generation.

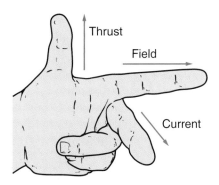

SINE WAVE GENERATION AND CONVERSION TO DC

Look at the **armature** loop at position A at the top of Figure 13–7. The position of the wire loop is such that it is moving parallel to the magnetic lines of flux. Now look at the waveform at position A. When the wire loop motion is parallel to the magnetic lines of flux, there is no voltage being produced because the wire loop that makes up the armature is not cutting through any of the lines of flux when the loop is parallel to the lines of flux.

FIGURE 13-7 Generation of sine wave by rotating a conductor through a magnetic field.

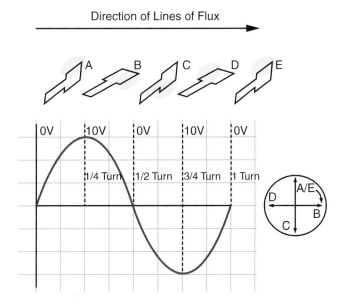

Armature

In a DC generator, the electrical part of the generator is connected to the load through the output connections to the armature. The armature is the rotating conductor as represented by Figures 13-2 through 13-5. The rotating armature is also mechanically connected to the prime mover, or mechanical power input. In AC generators, called alternators, the armature is connected to the output load but it is not always the moving part of the alternator.

Now rotate the armature from position A to position B. As the armature rotates, the motion of the wire loop goes from being parallel to being perpendicular to the lines of flux. As the armature rotates through this 90° travel, it begins to cut through more and more lines of flux.

When the armature gets to the 90° point, it is cutting the maximum number of lines of flux per second. Now look at the waveform at position B. Note that when the armature is cutting through the most lines of flux, the highest amount of voltage is produced in the positive direction.

Now rotate the armature through another 90°. At this point, the armature is 180° from its original starting point. Note again that the armature is moving parallel to the magnetic lines of flux and that the voltage produced is zero.

As the armature continues to rotate through the next 90°, the armature is again approaching the point where it will be moving perpendicular to the lines of flux. Take a look at the waveform and notice that the value has gone negative. This change in polarity is due to the motion of the leading edge of the wire loop relative to the direction of the magnetic lines of flux. Notice that the leading edge of the wire loop is indicated by a yellow highlight. When the leading edge is moving up, the polarity of the voltage being produced is negative.

As the loop continues to rotate back to its starting point, the voltage again begins to approach zero. The voltage being produced as this wire loop (the armature of the generator) rotates through a complete 360° of travel through the magnetic lines of flux is called alternating voltage because the voltage travels positive for a part of the cycle and then alternates and becomes negative for part of the cycle.

Since the voltage produced by all rotating armatures is alternating voltage, as can be seen in Figure 13–7, the method of removing the voltage from the armature is the key to determining whether the output voltage of the generator is alternating or direct.

13.1 Commutating the Output of an AC Armature

Since DC generators must produce DC voltage and resultant current, the method of removing the voltage from the armature is different than that used for an AC generator. In the DC generator, the components used to remove the voltage from the armature are brushes and a **commutator** (Figure 13–8A).

Since the commutator is physically connected to the armature loop, the segments have an AC voltage on them. However, in the example generator in Figure 13–8A, the commutator is positioned so that the segment that is shown on top, is always the negative side of the circuit. This means that the brush on the top will always be the negative brush, and the bottom brush will always be positive. The output voltage from the brushes will be a pulsating DC going from zero voltage to peak voltage but never reversing (Figure 13–8B).

Unlike Figure 13–7, which shows the voltage produced as being AC, Figure 13–8B shows the resulting waveform of the voltage when the voltage is measured at the brushes. Notice how the output is always positive, but pulsating DC. It is not smooth like a battery voltage but is DC.

As the armature loop moves from position A to position B in Figure 13–8B, a positive voltage is produced that starts at zero and gradually increases to maximum. As the rotation continues from position B to position C, the loop continues to cut through the lines of flux; however, as the armature position approaches position C, it is producing less positive voltage. When the armature reaches position C, it is moving parallel to the lines of flux, and therefore no voltage is produced.

From position A through position C, the loop produces a positive voltage that has been removed from the armature windings by the **brushes** that have been riding on the same commutator segments throughout this 180° of

Commutator
A multi-segment rotating connection that is connected to the armature windings. A commutator is a circular ring that is divided into segments that are electrically insulated from each other.

Brushes
Sliding contacts usually made of a carbon or graphite alloy that are positioned so they are always connected to the same polarity armature segments. The brushes ensure that the output from the AC voltage created on the armature is commutated (rectified) to DC.

FIGURE 13–8   (A) A rotating armature with commutator connections to produce DC. (B) The output from a DC commutator is a pulsating DC voltage.

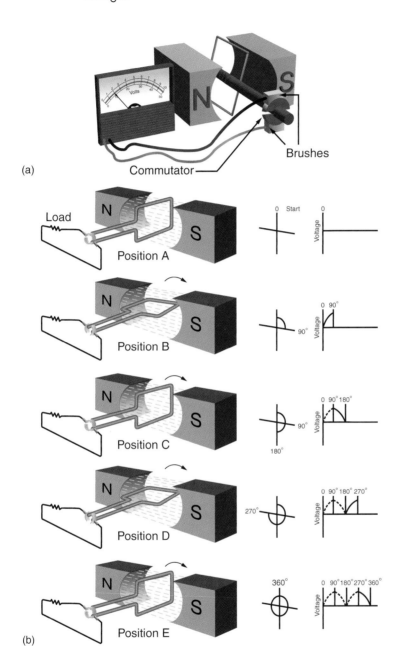

armature travel. However, as the armature begins its travel from position C to position D, the brushes will begin to ride on the opposite commutator segments. The change in commutator segments will allow the voltage to remain positive even though the armature is producing an alternating voltage.

As the commutator rotates from position C to position D, the positive lead of the generator is now connected to the segment of the commutator that is producing a positive voltage. At position D, the armature loop will again produce its maximum voltage and then return to zero as the armature rotates to position E.

Regardless of which half of the loop is producing a positive voltage and which half is producing a negative voltage, the commutator and brushes will always cause the output voltage to remain at the same polarity. This is the basic principle of DC generator operation.

MAJOR GENERATOR COMPONENTS

The rotating member of the DC generator is the armature. The entire assembly includes the iron core, the commutator, and the windings. There are three basic types of windings, depending on the needs of the machine: lap, wave, and frogleg (Figure 13–9).

Because each loop of the coiled wire in the armature has voltage induced into it, the additional windings on the armature will add to the total generated output. If the coils are connected in parallel with each other, the voltage is not additive but the current output is additive. If more voltage is desired, the coils can be connected in series in such a way as to have the voltages added and not the current. A combination of series and parallel-connected windings creates a higher voltage and higher current. These three conditions can be visualized in the three common connections such as the lap winding where the windings are in parallel, the wave winding, which illustrates the series connection to increase voltage, and the frogleg pattern, which increases voltage and current capabilities. See Figure 13–9 and the following descriptions.

FIGURE 13-9   Schematic diagrams of three types of armature windings.

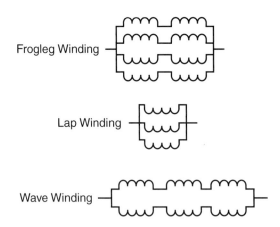

Frogleg Winding

Lap Winding

Wave Winding

The connections to the armature and commutator are through brushes. The armature connection points are labeled A1 and A2.

1. Frogleg armatures are the middle-of-the-road type of windings. The frogleg configuration, the most common of the winding configurations, is designed

for moderate current and moderate voltage. The frogleg armature, commonly found in larger DC equipment, has its winding connected in series and parallel. That is, there are groups of series-connected windings, and these groups are then connected in parallel.

2. Lap-wound armatures are used in machines designed for low voltage and high current. Typically, lap windings are designed with large wire to safely handle large currents. When equipment has lap-wound construction, the windings are connected in parallel. This parallel path construction provides multiple paths for current flow. A common characteristic of lap-wound equipment is that there are as many pairs of brushes as there are pairs of poles. Another characteristic of lap-wound equipment is that there are as many armature current paths as there are pole pieces.

3. Wave-wound armatures are used in machines designed for high voltage and low current. Since the current capacity of wave-wound equipment is relatively low, the winding wire size is smaller than that of the lap wound and is typically connected in series so that the voltage of each of the windings is additive. Keep in mind that in series circuits, the current is the same throughout the circuit. The same is true with wave-wound windings. Even though the windings are connected in series to increase the voltage, the current remains the same throughout all the windings.

13.2 Brushes

Brushes ride on the commutator to connect the rotating armature to the load (Figure 13–10). The brushes are normally made from a carbon-based material. Carbon is used for three reasons:

1. Carbon is softer than copper and allows the brushes rather than the commutator to wear (the latter being more difficult to repair).

2. Carbon withstands the high temperatures that may be present because of arcing and friction.

3. Carbon, copper, and moisture form a very thin, conductive film on the commutator that lubricates and reduces wear.

FIGURE 13–10   Photo of carbon brushes riding on a DC commutator.

13.3 Field Windings

The field of the generator produces the magnetic field for the armature. In a spinning armature, the field windings are on the stationary piece, or the stator. The field is connected to a DC source of power and the magnetic field produced by the coiled wires creates the flux for the magnetic field needed for generation by induction.

There are two types of **field windings** (Figure 13–11): the series field and the shunt field. Series field windings are connected in series with the armature through the brushes. Series windings consist of very few turns and very large wire. The terminal leads of the series field are labeled S1 and S2.

Field windings
The part of the DC generator that creates the magnetic field that is cut by the armature windings.

> **FIGURE 13–11**   Stator field winding of series field and separate shunt field.

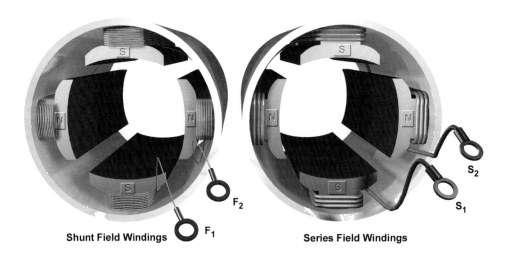

Shunt Field Windings F_1 F_2

Series Field Windings S_2 S_1

The shunt field is connected in parallel (shunt) with the armature. They are made with many turns of a very small wire and therefore have a higher resistance than a series field. The shunt field is also called the field and is labeled F1 and F2. The purpose of the field is to produce magnetic flux. The more magnetic flux a particular generator has, the more output voltage it produces. There are controls and limits to the amount of flux available in a magnetic field winding. The amount of current that is allowed to flow in the winding will directly affect the flux — the more current, the more flux— until the iron is magnetically saturated and cannot concentrate more flux lines. Generally the magnetic flux is controlled within the operating range of the magnetic iron, by adjusting the field current to the windings.

13.4 Pole Pieces

Mounted inside the housing of the generator are pieces of metal, usually made of soft, laminated iron or some other easily magnetized metal. These are fixed pieces and do not rotate. The purpose of the pole pieces is to concentrate the magnetic field. The field windings are wound around these pole pieces, forming an electromagnet. The electromagnet sets up the magnetic field necessary for generation. In small generators, the pole pieces are sometimes permanent magnets (Figure 13–12).

> **FIGURE 13–12**

Photo of permanent magnets used for field permanently mounted in stator.

GENERATOR CONNECTION TYPES

13.5 Series Generators

As the name implies the series generator is connected so that the series field windings are connected in series with load to the armature. As current flows to the connected load it passes through the series field and the generator voltage increases. The series generator must be **self-excited,** which means that the pole pieces must contain some amount of residual magnetism to start producing a voltage in the armature. In other words, the magnetism left in the iron stator pole pieces must be strong enough to start producing output voltage. As the output load current and the series field current increase, the magnetic field of the generator becomes stronger and further increases the output voltage. The process continues as shown in Figure 13–13 until the controls on the field excitation won't allow the field to become any stronger, or the iron core becomes magnetically saturated. The residual magnetism produces an initial output voltage that permits current to flow through the field if a load is connected to the generator. In a series generator, the output voltage increases as the load current increases.

Self-excited generator
Self-excited refers to the method in which the generator receives power to produce the magnetic field for generation. A portion of the output power is fed back to the generator's field as it provides its own **excitation current**.

Excitation current
The current supplied to the field of an AC generator. The excitation current creates the steady magnetic field that the armature cuts through.

FIGURE 13–13 Output voltage rises with increase in excitation until saturation of iron prevents any further increase in induced voltage.

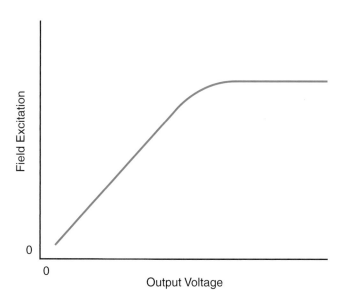

As additional loads (current paths) are connected to a series generator, the output voltage will continue to increase until the magnetic cores saturate (Figure 13–14) or the field control resistance (diverter) prevents any additional current from going through the series field. The amount of voltage generated by a series generator is dependent on the strength of the magnetic field produced by the pole pieces, the number of turns of wire on the armature, and the speed of rotation of the armature. The more field current, the more voltage; the more speed the more voltage; the more turns of the coils on the armature the

FIGURE 13–14   A series generator will generate higher voltage until field
saturation or the diverter "diverts" current from the series field.

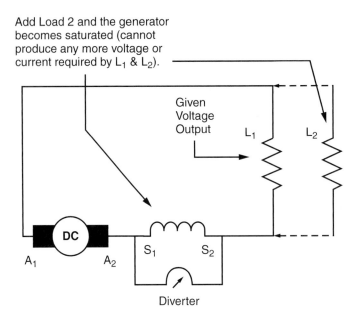

Add Load 2 and the generator
becomes saturated (cannot
produce any more voltage or
current required by L_1 & L_2).

Given
Voltage
Output

L_1 L_2

DC

A_1 A_2 S_1 S_2

Diverter

more voltage. Because of the varying voltage, series generators are best suited for use with a constant load.

13.6 Shunt Generators

Shunt generators, the second winding configuration for DC generators, get their name from the field windings being wired in parallel with or shunted across the generator output. Self-excited shunt generators make use of residual magnetism to start the generation process. This is done by applying the residual voltage (generated by the residual magnetic field left over from the last operation) to the shunt field to create an initial current. The shunt generator provides maximum output voltage before the load is applied. As the load increases on a shunt generator, the output voltage will decrease.

To attempt to control the output voltage of a shunt generator, an electronic voltage regulator is sometimes installed as part of the generator circuit. The electronic voltage regulator senses the changes in the output voltage and makes adjustments to the shunt field current. The regulator causes less current to flow through the shunt circuit when the voltage is sensed to be too high, or more current to flow in the shunt field when the output voltage goes too low. Figure 13–15 shows how the voltage regulator, the shunt field, and the output voltage are all related. A simplified explanation of the voltage regulator box is that it acts like a series variable resistor (or rheostat) and voltage divider to the shunt field. This will regulate the output voltage available to the load. In reality, the device is more than a series resistor. The circuitry is much more complex in that it has a controller (sometimes a microprocessor-based computer).

FIGURE 13-15   A regulator can be used to adjust the amount of current through the shunt field thereby controlling the output voltage.

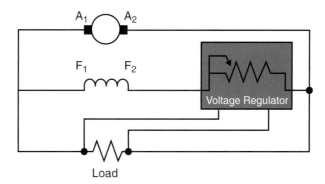

13.7 Compound Generators

As we have seen, the generator voltage can change from the no-load condition with no current output to the full-load condition with full-rated current output. In most cases, the equipment connected to the generator should have the same voltage output whether at no-load, light-load, or full-load. In order to provide this feature, some adaptations to the series generator of the shunt generator is required. We can combine the best features of each generator and create a compound generator.

Compound generators contain both series and shunt fields (Figure 13–16). The relationship of the strengths of the two fields in a generator determines the amount of compounding for the machine. Since the voltage regulation of a series generator can be very poor and a shunt generator has good voltage regulation, combining the two gives better circuit flexibility and load control.

FIGURE 13-16   Schematic of "long-shunt" connection for compound wound DC generator.

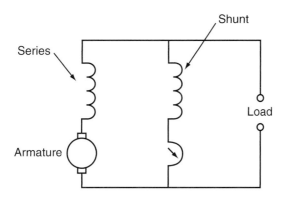

There are two subdivisions of the compound generator: short-shunt and long-shunt, compounding. The long-shunt compound variation has the shunt winding into the *longer* connection point in parallel with the combined armature and series winding circuit.

The short-shunt configuration has the shunt field connected in parallel with the armature, or *short* connected across the armature. The series field is then wired in series with the armature and the load (Figure 13–17).

FIGURE 13–17 Schematic of "short-shunt" connection for compound wound DC generator.

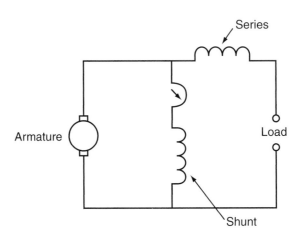

Flat compounding
The effects of compounding in a DC generator so that the "no load" output voltage is the same as the "full load" voltage. The voltage vs. load curve on a graph is essentially flat, or no change in voltage, with change in load.

Overcompounding
A condition that occurs is a DC generator when the output voltage at "no load" is less than the output voltage at "full load." The output voltage increases as electrical load is added.

Undercompounding
The condition that exists when the setup of the generator results in the "no load" voltage being greater than the "full load" voltage. As the generator supplies more load current, the output voltage decreases.

Depending on the number of turns in the series field vs. the shunt field and the controls that determine the current through the two fields, there are three possible ways (classifications) of compounding windings in a generator: **flat compounding, overcompounding,** and **undercompounding.** Figure 13–18 shows the typical voltage and current relationships in the different compounds. When the compounding of series and shunt fields aid each other in producing electromagnetic force (inducing voltage and current), the compound is called cumulative compounding. When the series windings oppose the shunt field windings in producing electromagnetic force, the compound is called differential compounding.

Refer to Figure 13–18. The following are characteristics of the different compound generators:

1. The overcompound generator's full-load voltage is greater than the no-load voltage with cumulative compounding.

2. The undercompound generator's full-load voltage is less than the no-load voltage, but the series component of the compound makes it a better voltage regulator for loads than the shunt. This is still a cumulative compounded generator.

3. The flat-compound generator has a load voltage equal to the no-load voltage and is still cumulative compounding connection.

4. The differential compound generator output voltage drops as more current is drawn by the load. Differentially compounded generators were used for DC welding purposes.

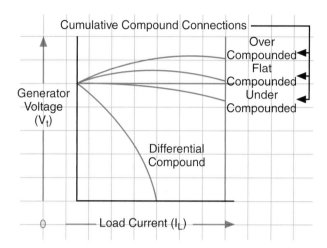

FIGURE 13–18   Voltage output curves for different connections of generators under load.

Most commercial compound DC generators are normally supplied by the manufacturer as overcompound machines.

GENERATOR LOSSES

The power equation for a generator or motor is $P_{in} = P_{out} + P_{loss}$. This means that the power used to turn or rotate a generator must always be greater than the output power delivered by the generator to perform work. It also means that not all the power a generator receives can be converted into useful mechanical or electrical energy; there is a loss, and this loss is almost entirely heat. The relationship between power out and power in is called the generator's efficiency, and is noted with the lowercase Greek eta (η). The equation for efficiency of a machine is:

$$\eta = \frac{P_{out}}{P_{in}} \quad or \quad \% \text{ Efficiency} = \left(\frac{P_{out}}{P_{in}}\right) \times 100$$

For a generator, the equation becomes:

$$\eta = \frac{P_{out}}{(P_{out} + P_{loss})}$$

And for a motor it is:

$$\eta = \frac{(P_{in} - P_{loss})}{P_{in}}$$

Generator power losses can be broken into two main classes: electrical and mechanical—(friction and windage). Electrical losses are caused by current flow through various parts of the generator and are sometimes referred to as copper losses. The greatest of these losses is due to the resistance of the windings. The watts loss could be calculated as $I^2 \times R =$ watts loss.

TechTip!

FIGURE 13–19

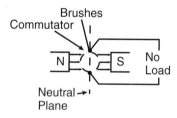

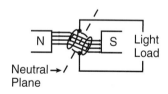

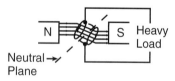

The power loss due to induced currents in the iron core material of a generator is called eddy current loss. This is induced as the armature spins through the flux lines of the pole pieces field. Eddy currents are electrical currents that circulate within the metal. As explained in Chapter 1, the currents can circulate within the conductor. The magnetic field that is cutting the conductor and the iron core induces current into the iron. This induced current is small, but the electric currents tend to circulate in the iron as "eddies" of current. Because the electrical resistance of the iron is high, the currents are small, yet the effect is to create a watt loss in the form of $I_2 \times R$ losses. These losses due to eddy currents in the iron are referred to as iron losses. By cutting the iron into thin slices, and putting them back together, or **laminating** the iron, the resistance of the iron goes up and the eddy currents go down, thereby reducing the iron losses. The iron losses also consist of the hysteresis losses or friction of the molecules in the iron as they tend to align themselves in the magnetic field. Hysteresis is explained in detail in the study of DC Theory.

As the DC generator rotates, an alternating current is set up in the armature. This causes the molecules of the iron to realign each time the current changes direction from negative to positive which changes the magnetic field. This molecular friction produces losses due to hysteresis. Hysteresis losses can be varied by using different grades of steel for the rotor core. High hysteresis steel would have higher losses in the magnetizing of the iron which are characterized by heat. To reduce the heat (energy) loss due to hysteresis, high silicon steel can be used. Together the iron losses, the copper losses, and the mechanical friction losses of the bearings along with the windage losses as the fan moves air over the coils, all add up to the generator losses.

ARMATURE REACTION

In a generator, there are two primary magnetic flux fields: the flux field produced by the field windings around the N and S poles and the armature flux field. Look at Figure 13–20. The current flow in the armature conductors produces its own magnetic field, which reacts with the stator magnetic field. The larger the armature current (the more current to the load) the more magnetic affect it has on the stator, or main magnetic field. The rotor reacts with the stator field to bend the stator field. Notice that if the armature's magnetic field (flux) shifts the magnetic neutral, the magnetic flux field for the field windings becomes distorted. This distortion of the magnetic fields is referred to as **armature reaction.** It can cause a heavy circulating current to be produced in the armature that will cause arcing between the commutator segments and the brushes. The arcing between the commutator segments is referred to as commutation. This happens because the amount of voltage that is induced in the armature windings is not balanced at a neutral physical location. The neutral point referred to as the neutral plane, where the brushes are located, normally has a balanced amount of voltage on each side of the neutral plane. As the neutral plane shifts because of field distortion, the voltages are no longer balanced. This is explained in more detail while studying motors and generators.

Armature reaction
The bending of the field magnetic flux by a reaction to the armature magnetic field is known as armature reaction. Armature reaction causes a displacement of the neutral point with consequent arcing at the commutator and resultant voltage drop.

FIGURE 13–20 The twisting of the magnetic field because of current in the armature creating an armature reaction.

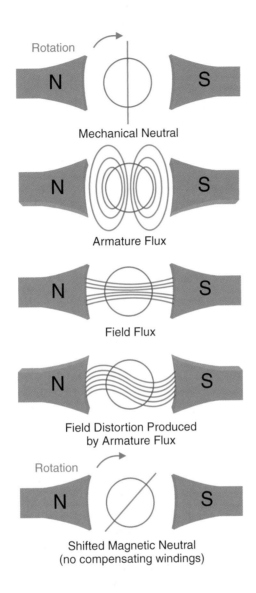

Mechanical Neutral

Armature Flux

Field Flux

Field Distortion Produced
by Armature Flux

Shifted Magnetic Neutral
(no compensating windings)

To offset the effect of the field distortion from the armature's shift caused by the armature's own field from the magnetic neutral plane (caused by self-induction), compensating windings are used to produce a counter-electromagnetic force (Figure 13–21). These windings are placed between the main field poles and are called by several names such as **compensating, interpole,** or **Thomson-Ryan** windings. The current to the load causes a magnetic field in the compensating windings that is proportional to the load. The more current to the load, the more the rotor magnetic field tends to bend the main magnetic field. The compensating windings oppose the bending efforts and, therefore, keep the neutral plane straight even with a variable load.

Compensating, interpole, or Thomson-Ryan windings
Compensating windings, interpole windings, or Thomson-Ryan windings are field windings that are located physically between the main field poles. Interpoles are used to reduce armature reaction and are connected in series with armature current.

FIGURE 13-21 Location and effect of compensating windings on the neutral plane of the generator.

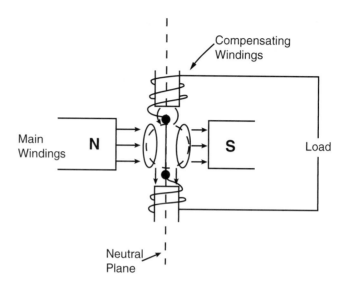

FIGURE 13-22

Paralleling DC compound wound generators requires the addition of an "equalizer" connection.

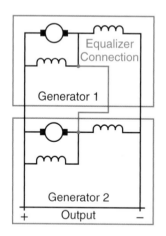

Slip rings

Continuous bands of metal installed around a shaft. Slip rings are connected to the rotating windings and provide a path for the current to reach the brushes (see Figure 13-23).

PARALLELING DC GENERATORS

There may be a need for connecting DC generators in parallel to supply enough current for a connected load. An equalizer connection is required to prevent one generator from taking the load and the other from acting as a DC motor. Look at Figure 13–22 to see a schematic of an equalizer connection. The equalizing connection is used to connect the series fields of the two machines in parallel with each other. This parallel connection maintains the same voltage across all series fields and prevents one machine from taking over the other machine as a motor. Notice that the connection must be on the armature side of the series field and connected to the same polarity on each generator. The two generators must be connected in the same pattern, designed for the same voltage output, and both must have "drooping" speed characteristics. This means that they must act the same way as electrical load is added in that they slow as they do more electrical work. As you study generators and motors in more detail, these characteristics will become more apparent.

AC GENERATOR PRINCIPLES OF OPERATION

The generation of the voltage is the same basic process used in the DC generator (Figures 13–23 and 13–24). The simplest AC generator is constructed of a permanent magnet, **slip rings** (instead of a commutator), brushes, and a single loop (armature). A voltmeter to read generated voltage is the external load. The armature is the rotating part of the machine in which the voltage is induced exactly the same as the DC description. The AC description of generating an AC voltage and resultant current is detailed in Chapter 2. To quickly recap the process use the following summary: The magnets set up the required magnetic field flux. Each end of the wire loop is connected to a separate slip ring. As the loop rotates, each side of the loop will be cutting

FIGURE 13-23   Generation of AC voltage with rotating armature and slip rings.

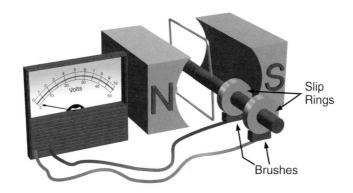

FIGURE 13-24   Steps of AC generation and output waveform.

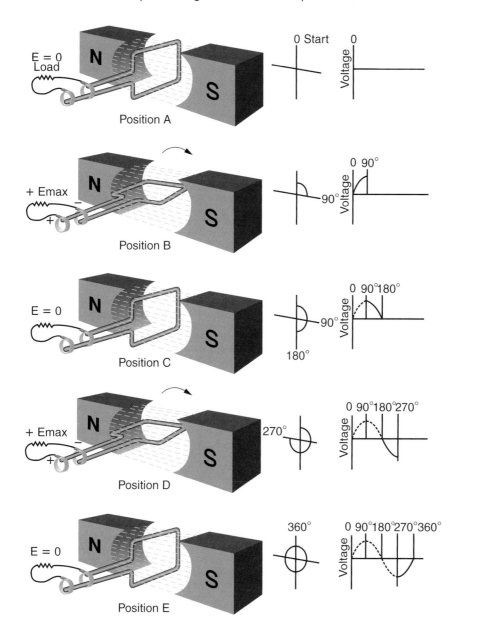

lines of flux at the same time. As one side of the loop moves downward, the other side moves upward.

In position A in Figure 13–24, the loop is rotating, but it is cutting very few lines of flux. This position is called the 0° position. If the voltage were being measured here, it would be zero (see Figure 13–24, 0 V).

In position B in Figure 13–24, the generator has rotated 90°, and both sides of the loop are cutting through the lines of flux perpendicularly. At this point, the generator is producing the highest voltage. If the voltage were being measured, it would show the output of the generator to be at its highest "positive" value (+10 V). Notice also that the output increases gradually as the loop rotates from 0 to 90°.

In position C in Figure 13–24, the generator is producing the least amount of voltage again, but this time the loop is in the 180° position, opposite the position shown in position A. The zero voltage position is caused by the generator coil traveling parallel to the flux lines (not cutting through them) and this position can be referred to as the zero-crossing point..

Position D in Figure 13–24 shows the loop in the 270° position. It is now cutting the maximum number of flux lines and producing the maximum voltage. The voltage is now "negative" when compared to the voltage being produced in position B. This is because the current is now in the opposite direction through the loop (−10 V).

CONSTRUCTION OF AC GENERATORS

AC generators, also known as alternators, operate on the same principle of electromagnetic induction as DC generators. Both types of generators pass conductors through a magnetic field to create a voltage and resultant current flow. However, unlike the DC generator, which uses a commutator to rectify the generated voltage, the AC generator uses slip rings to connect the AC output directly to the load. The different outputs of DC and AC generators can be seen in Figure 13–25.

FIGURE 13–25   The output voltage from the commutator of a DC generator with multiple poles may look like the top figure. The bottom figure is an alternator output taken from slip rings.

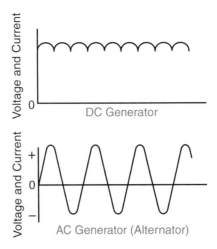

There are two basic types of alternators: revolving armature and revolving field. The revolving armature is used for most basic explanations as it is easier to visualize. However, the revolving armature is used primarily for small-load applications. Both revolving armature and the more prevalent revolving field AC generators produce an alternating sine wave output voltage.

13.8 Revolving Armature AC Generators

The **revolving armature** AC generator is not as common as the revolving field generator. It is similar in construction to the DC generator in that both types use some form of mechanical moving contacts to connect the external circuit to the armature circuit. The AC generator uses slip rings instead of a commutator. The purpose of the slip rings is to directly connect the armature windings to the load. Since no commutator is used, the output is an alternating voltage instead of a direct voltage.

Revolving armature generator A generator that has the field windings on the stator and the armature windings on the rotor.

The slip ring, unlike the segmented DC generator commutator, is constructed as a continuous ring and is not broken up into isolated segments. Each individual end of the rotating loop of the armature is connected to a separate slip ring (Figure 13–26).

FIGURE 13–26 Single phase generation with a revolving armature.

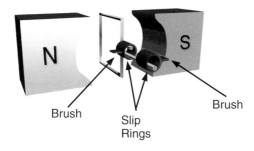

In a three-phase revolving armature AC generator, there are three separate sets of slip rings, one for each phase. Each ring would be connected to a windings end. The windings, or loops, are electrically oriented 120° from each other (Figure 13–27). One end of each coil is connected together on the rotor and three slip rings are used to bring the three coil ends to an external circuit. In this case, the generator is connected in a wye pattern. This is a typical connection. Wye connections for three-phase systems will be explained in detail in Chapters 14 and 15. The output of the three-phase generator would resemble the output wave, as seen in Figure 13–28. As you view Figure 13-28, you will notice that the peak of each phase is 120-electrical degrees away from the next peak. This is true for three-phase AC generators. We can connect the generator to a three-phase power system with three-line wires representing phase A–B–C as commonly denoted.

FIGURE 13-27 Three phase generation with 3 coils spaced 120° apart connected in wye pattern.

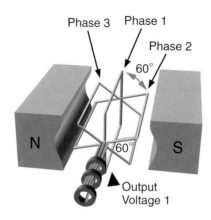

FIGURE 13-28 Graphic representation of voltage versus time showing three phase voltage production.

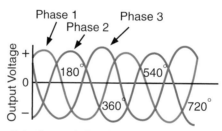

Note: One end of each generator (alternator) loop is connected together.

Although the revolving armature AC generator can produce an alternating output voltage, the slip rings and brushes limit the size of load it can carry. The higher currents require large slip rings and large brushes and highly increased maintenance. As the currents and voltages become higher, the sliding contacts are not as efficient and the arcing under the brushes creates heat and wear, and the voltage drops. The high voltages required by some loads mean that much heavier insulation on the armature conductors must be used, which makes the armature larger and heavier to spin. Taken together, these features limit the kilovolt-ampere (kVA) capacity of the revolving armature generator.

13.9 Revolving Field AC Generators

Revolving field
A generator that has the field windings on the rotor and the armature windings on the stator.

The **revolving field** type of generator has the armature windings on the stator and the field windings on the rotor. By mounting the windings that provide the output voltage in a stationary position and rotating the magnetic field, the capacity and connection limitations experienced with the revolving armature type of AC generator are eliminated. Remember that we only need relative motion between the magnetic field and the armature coils. It does not matter whether the conductors move through the magnetic field or the magnetic field cuts through the stationary conductor. Higher voltage and kVA ratings are possible with the revolving field AC generator because the output circuit can be hardwired through stationary, secure connections to the stationary armature, instead of being connected through a brush and slip ring arrangement.

The portion of the generator that rotates is the rotor and in this case holds the DC magnetic field. The stationary portion is the **stator** which is wound with coils to create an armature. Figure 13–29 shows an example of a three-phase revolving field generator. This type of generator stator is constructed by placing the three sets of windings 120° apart. As you can see, there is a winding that belongs to each phase on each side of the stator. Again, they all have a common end tied together in the stator as you note in the connection point under the bottom coil.

Stator
The stationary part of the generator.

TechTip!

The actual voltage generated in the coils may be measured as 120 V RMS from one coil end to the other. When we measure the voltage from one output lead to another, we would read 1.73 times the actual coil voltage; in this example 120 V RMS × 1.73 = 208 V RMS. This increase in voltage is a result of the phase voltage being out of phase by 120 electrical degrees. Again, we will add voltages vectorially to yield the higher voltage. This is an automatic increase in the generated voltage of 120 V per coil to a higher line voltage when the windings are in a wye connection. We will use this effect to our advantage as we transmit and use the three-phase power from the generator.

FIGURE 13–29   Three phases mounted on the stator of a rotating field generator.

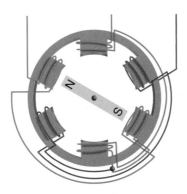

3 Phases, 120° Apart

ROTOR CONSTRUCTION

Rotating fields are constructed in two different ways: the salient pole rotor and the round rotor. The round rotor is also called a turbo rotor.

13.10 Salient Pole Construction

Figure 13–30 shows the **salient pole** construction. This particular rotor is one that is used on a very large hydroelectric generator. As you can see there are many individual, or distinct, poles around the periphery of the rotor. This is because the number of poles and the speed of the rotor will dictate the output frequency. With hydroelectric generators, the speed is quite slow, so the number of magnetic poles is high. Much smaller salient pole rotors are used in some transportable generators.

Salient pole
Salient means projecting outward; projecting or jutting beyond a line or surface; protruding. A salient pole rotor is a type of rotating field construction where the field poles are wound individually and mounted along the outside edge of the rotor.

FIGURE 13-30   The rotor of a rotating field generator using "salient pole" construction.

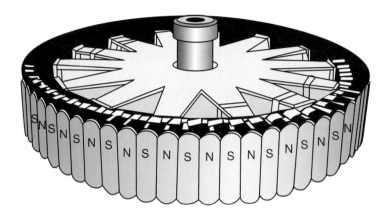

Notice the pole faces arranged around the outside of the rotor. They jut out and are very pronounced. This is why they are called salient poles. Each one of the pole faces will be magnetized by its own set of coils. North and south poles are adjacent to each other so that every north pole is between two south poles and vice versa.

This type of rotor construction is very heavy because of the iron pole faces and the copper windings. Such rotors are subject to very high wind resistance (making it harder to spin) and centrifugal forces.

13.11 Round Rotor Construction

Round rotor

A more streamlined type of construction where the coils are wound longitudinally on the rotor. This type of construction is generally much lighter and is used extensively on high-speed generators. Also called a turbo rotor.

As you can see in Figure 13–31, the **round rotor** has windings that are wound longitudinally. Round rotors are generally smaller and lighter and are found in the high-speed, two- or four-pole generators used in large utility generating stations. Although such a rotor usually turns much faster than the salient pole rotor, its longer, narrower construction minimizes wind resistance and centrifugal forces. The outside surface of the rotor is smooth and the actual field windings are not exposed.

FIGURE 13-31   A round rotor of a rotating field generator where specific poles are not visible.

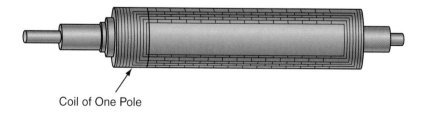

Coil of One Pole

EXCITATION SYSTEMS

Regardless of whether the field is revolving or stationary, it must create a magnetic field so that the armature windings can cut through it (or be cut through, in the case of the revolving field) and produce the output voltage. Many different types of excitation systems are used. A few of the more important ones are discussed here.

13.12 Permanent Magnet Fields

Some smaller generators use a permanent magnet to provide the field. This type of construction is usually limited to small auxiliary generators used for excitation or control purposes. Such a generator is called a **PMG (permanent magnet generator)** (Figure 13–32). In this generator there is no need for a DC source of power to create an excited magnetic field. The speed is set to produce the desired frequency and the output voltage is controlled external to the generator.

PMG
Abbreviation for "permanent magnet generator." In this type of generator, the field is provided by a very strong permanent magnet sometimes in the stator, sometimes on the rotor.

FIGURE 13–32   The stator of a PMG generator with permanent magnets on the stationary portion.

13.13 DC Electromagnetic Field

Most generators create a magnetic field (magnetic field flux) on the rotor or stator by using DC to "excite" the coils of wire, creating a steady magnetic field. Generators use some sort of DC supply to provide the current for their field excitation. The DC source that provides this current is called an exciter and the current it provides is called excitation current. Exciters are made in a variety of different ways, including the fixed exciter and the rotating exciter.

An example of a fixed exciter for a revolving field generator is shown in Figure 13–33. Here a battery is connected to the windings of the field magnets mounted on the rotor. The battery supplies the DC excitation that creates the magnetic revolving field that cuts through the armature windings. The output of the armature is then connected to the load. In this case, the battery is the exciter for the generator, which is not a practical application because there is no control and the battery will discharge.

FIGURE 13-33 The excitation voltage for the DC field on the rotor shown supplied by a battery.

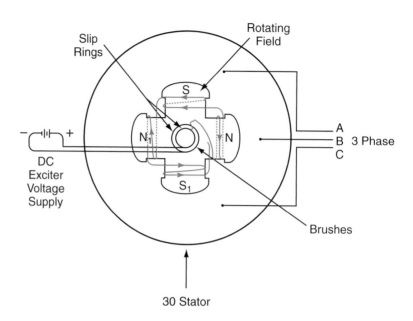

A block diagram of a fixed exciter is shown in Figure 13–34. In this type of exciter, the battery is replaced with an AC supply, a rectifier, and a DC voltage regulator. The transformer changes the voltage to a level suitable for the exciter input. The regulator controls the generator voltage output by regulating the amount of DC excitation current supplied to the field windings. A sample of the generator output voltage is fed back to the regulator as the reference for voltage control. Some form of this type of feedback from the actual output voltage is typical of an AC generator with voltage controls. As expected, if the output voltage decreases due to current load increase, the regulator senses this and in turn increases the exciter voltage and exciter current to produce a higher output voltage. The opposite is true for a voltage decrease as the regulator works to keep the output voltage steady.

FIGURE 13-34 A block diagram of a voltage regulator adjusting the DC excitation voltage to generator field.

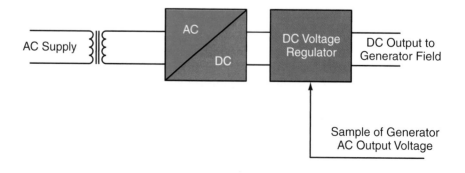

A rotating exciter is a separate DC generator that supplies the DC current to the field windings of the main generator. This type of exciter can be separate from the main generator or mounted on the same shaft as the main generator. Sometimes this generator is simply a DC generator driven off the same shaft as the AC generator. The output of the DC generator is controlled and fed to the exciter field on the alternator via slip rings and brushes. If it is mounted on the same rotating member as the alternator's exciter, it could be called a brushless exciter.

In a brushless exciter, the armature of the exciter is mounted on the same shaft as the field windings of the main generator. Since the two coils are rotating on the same shaft, there is no need for brushes or slip rings. However, since the output of the exciter armature is AC, it must be changed to DC. This is done by attaching a circular disk to the shaft. The disk has a diode rectifier mounted on it. The output from the exciter is connected to the input of the diode rectifier. The output of the diode rectifier is connected directly to the rotating field windings. The voltage level on the system is controlled by regulating the field current of the exciter.

GENERATOR COOLING

Generators with small kVA outputs are normally air-cooled. Appropriate openings are left in the stator windings, and slots are provided for air to pass through. Air-cooled alternators will usually have fan blades attached to one end of the rotor shaft that help circulate the air through the entire unit. Figure 13–35 shows a small generator cooling fan.

TechTip!

The brushless exciter makes use of diode rectifier circuits mounted on the rotating member of the AC generator. The rotating member has an AC voltage induced into its windings as the rotor spins past a stationary DC field (see diagram). The AC voltage in the rotating armature is then rectified by the rectifier devices and used to supply another coil on the rotor with DC voltage, creating a spinning DC field. This rotating DC field acts as the exciter for the stationary output armature. Therefore, the rotor has both an AC armature and a DC exciter operating on the rotor.

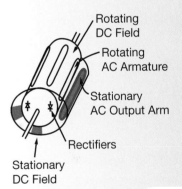

FIGURE 13–35   A fan is mounted to the shaft under the housing at the right. The fan blows air over the ribbed housing to cool the generator from waste heat.

Large-capacity alternators are often enclosed and operate in a hydrogen atmosphere. There are two reasons for using hydrogen as a "cooling" atmosphere. One is that the hydrogen is less dense, which means there is less wind resistance; therefore, less mechanical drag on the prime mover. The other is that the hydrogen atmosphere dissipates heat faster than air does so the heat from the copper losses and generator's AC losses ($I_2 R$, eddy current, and hysteresis) is removed and cooling is more effective.

FREQUENCY AND OUTPUT VOLTAGE

The frequency output of the voltage is a critical aspect of the AC generator's design. The output frequency is determined by the number of stator poles and the speed of rotation of the rotor. Because the number of stator poles is not field-changeable and is determined by the design of the generator, the only way to vary output frequency is by the speed of the rotor.

The formula for calculating the frequency when the poles and revolutions per minute (rpm) are known is:

$$f = \frac{S \times P}{120}$$

where:
    f = frequency in Hz
    P = number of poles
    S = speed in rpm
120 = a constant

The number of poles is the total number of north and south poles present in the generator stator. Since they always come in pairs, the number of poles will always be an even number. Two- and four-pole machines are the most common construction types for small generators, but many hydroelectric plants and other such generators may have hundreds of poles.

The constant 120 number is actually 2 × 60. The 2 changes the number of pole pairs to the number of poles, and the 60 changes rpm to revolutions per second to match the frequency in cycles per second or hertz. The full explanation of poles, speed, and frequency can be found in Chapter 2. We use the same formula for single-phase machines and three-phase machines. In three-phase machines, the number of poles is actually the number of poles per phase. For example in Figure 13–29, you see six magnetic poles— or two per phase. In the formula for frequency we would use two poles, so this would be a two-pole, three-phase machine. The frequency at 3600 rpm is 60 Hz.

Table 13–1 shows the relationship among frequency, speed of rotation, and number of poles for a variety of standard generator types. All the numbers in Table 13–1 were developed using the equation in the following example.

Example

What is the output of an alternator that contains four poles and is turning at a speed of 1500 rpm?

Solution:

$$f = \frac{(S \times P)}{120}$$

$$f = \frac{1500 \times 4}{120}$$

$$f = 50 \text{ Hz}$$

TABLE 13–1	Table Showing Relationship Between Speed, Number of Poles and Frequency		

	Desired Frequency		
Number of Poles	25 Hz	50 Hz	60 Hz
2	1500 rpm	3000 rpm	3600 rpm
4	750	1500	1800
6	500	1000	1200
8	375	750	900
10	300	600	720
12	250	500	600
14	214.3	428.6	514.3

OUTPUT VOLTAGE, FREQUENCY, AND POWER

Output voltage is just as important as frequency. Remember that the output voltage is a function of how many lines of flux are being cut each second. With that in mind, you can see that there are three generator parameters that will affect the voltage:

1. The length of the armature conductors, determined by the length of each turn and the total number of turns
2. The speed of rotation of the generator
3. The strength of the magnetic field

The number of turns of the coil is fixed by the generator's design, and the speed of rotation is fixed by the output frequency requirements. Thus, the only option for voltage control is to control the strength of the magnetic field. Since the excitation current controls the magnetic field's strength, it is used for the generator output voltage control. If the generator is acting as a sole source of power, the field control will determine the output voltage.

The amount of power (watts) that is produced is dependant on the input power. As there is more electrical current drawn from the generator, there is a force that slows the generator down or creates more mechanical load. This effect, called motorization or counter-torque of the generator, requires that there be more mechanical horsepower delivered to the input of the generator to keep the frequency constant. This automatic control is usually accomplished through a speed-monitoring system or a governor control on the mechanical input.

PARALLELING AC GENERATORS

There are three conditions that must be met when paralleling alternators:

1. The phase rotation of the two machines must be the same. This means that the A, B, and C phases must occur in the same sequence on both machines.
2. Each generator must have the same frequency (Figure 13–37).
3. The output voltage of the two machines must be the same.

TechTip!

As you have most likely observed at a carnival or fair with large portable truck-mounted generators, as the rides start and the demand for current increases, the generators have to work harder. This is usually apparent when the diesel engines work harder, creating more noise and more exhaust. As the electrical load or power requirement diminish, the engines settle back to the idle mode or throttle back. This is a direct indication that as there is more electrical work being done, more mechanical horse-power is required. Along with the mechanical changes are the voltage regulation controls, which are not as obvious but are constantly adjusting the magnetic field to maintain voltage.

FIGURE 13-36   Synchronized generator waveforms.

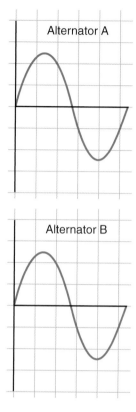

One of the most common methods of detecting when the phase rotation of one alternator is matched with that of a paralleled alternator is to use a phase rotation indicator as shown in Figure 13–37. This type of indicator will tell you whether the sequence of the phases is ABC or CBA. If the directions are not the same, there will be massive short circuits if you

FIGURE 13-37   A phase rotation indicator can be used to check the phase sequence of operating three-phase systems.

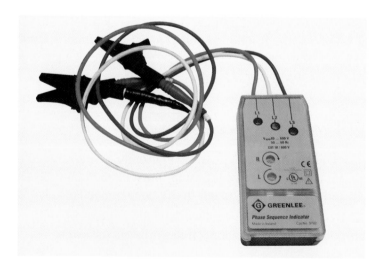

try to parallel the generators. Another method to use is the connection to three lamps. See Figure 13–38 for the lamp connection. Each lamp is powered by the voltage between the phases of each generator. The phase A voltage of generator 1 will connect to one side of a light, and the phase A voltage of generator 2 will connect to the other side. Likewise the lights are connected between B–B and C–C phases. When the two generators are in phase and the voltage magnitudes are equal, the lights will be off. This indicates synchronism, meaning the two generators are operating to produce a common voltage and frequency. This is the case of all generators on an AC power grid, or parallel generators used to power an emergency power source for a hospital.

FIGURE 13–38 A three lamp method can be used to indicate phase rotation and synchronism point.

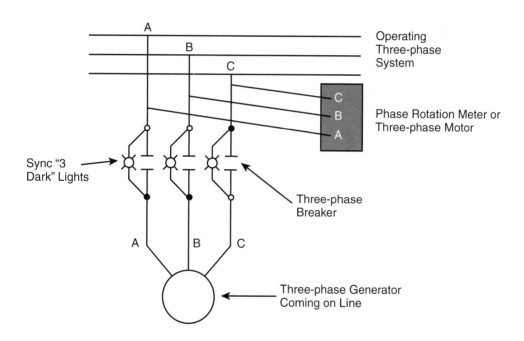

Another instrument that can be used to parallel two machines is called a **synchroscope** (Figure 13–39). This instrument measures the voltage differences of two alternators being placed in parallel. The base load alternator (the one previously connected to the load) is used as the reference point for measuring the phase voltage relation of the first alternator compared to the second alternator. Because the second generator is normally run slightly faster than the base generator, the synchroscope will register a "fast" reading. As the second generator exactly matches the peaks of the same phases, the synchroscope will be at the top dead center to indicate synchronism. The two generators are paralleled and the second generator will slow down as it assumes electrical load. Once paralleled, the machines will run together to maintain frequency. The power controls are more complicated and can best be studied in detail in specific generator books.

Synchroscope
A type of instrument that is used to determine the phase relationship between two separate generator voltages. Often used for determining synchronization of two generators prior to connecting the outputs together.

FIGURE 13-39 A phase rotation meter applied to two machines can verify same rotation sequence.

The synchroscope meter display will indicate whether there is a speed and therefore frequency mismatch between the two machines being paralleled. The electromechanical synchroscope looks somewhat like a clock with only one hand. The two voltages are connected to the inputs, and if they are in phase, the meter hand will point straight up (12 o'clock). The meter can rotate a full 360° in a clockwise direction. When the generators are in parallel there is no difference between the two generators and the needle stays at the 12:00 position.

SUMMARY

DC current has some applications that make it superior to AC. Many industrial plants use DC generators to produce the power needed to operate large DC motors or for needed DC power supplies.

Although most generators are AC, there are still some applications for DC. AC generators use slip rings to remove the voltage from the armature; DC generators use commutator segments. The commutator and brushes change the AC produced in the armature into DC output voltage. The brushes are used to make contact with the commutator and carry the output current to the outside circuit.

Generator markings include the following:

1. The armature connections are marked A1 and A2
2. The series field windings are marked S1 and S2
3. The shunt field windings are marked F1 and F2

The three factors determining the output of the generator are the following:

1. The number of turns of wire in the armature
2. The strength of the magnetic field
3. The speed of the armature

There are two basic types of AC generators (alternators) used to convert mechanical energy of a drive shaft to electrical energy. The two classifications for alternators are the revolving armature and the revolving field. The revolving armature is least used because of the limited power and voltage rating caused by brush and slip ring limitations when connecting the output load to the armature.

The rotor of the revolving field alternator is the rotating magnetic field. Direct current supplied to the field is called excitation current. The number of poles and the speed of the rotor determine the output frequency. The output voltage is controlled by the amount of DC excitation current to the rotating field, once the synchronous speed is set to maintain the desired frequency. This increases or decreases the strength of the rotor's magnetic field and results in induction of more or less output voltage and current. The only limit is the amount of current the conductors can carry based on the wire size, and when the iron core becomes magnetically saturated and cannot produce more magnetic effect even if the excitation is increased.

AC and DC generators can be operated in parallel to facilitate the transfer of load in order to take one generator off line for maintenance or to provide more capacity for an increasing load. Essentially all alternators that supply the nation's power grid are operating in parallel. This means that they need to be added or dropped off as the system's load changes. Proper paralleling of generators is often done automatically through electronic controls, but is sometimes required to be done manually as conditions dictate.

REVIEW QUESTIONS

1. Define a generator.
2. What type of voltage is produced in all rotating armatures?
3. In a DC generator, AC is converted to DC by the _____ and _____.
4. Name three types of armature windings connections and discuss how they are used.
5. Discuss the three factors that determine the voltage output of a DC generator.
6. Discuss the voltage regulation characteristics of the three different DC generator connections.
7. What advantages does a compound-wound field connection generator have over the shunt field connection and the series field connection?
8. What are the effects of armature reaction? What causes it? How can these effects be corrected at manufacture?
9. Explain how electricity is produced through electromagnetic induction.
10. If you have a wire and a magnetic field, how is voltage created on the wire?
11. On a DC generator, how do you reverse the polarity of the voltage to the load?
12. How do you increase or decrease the amplitude of the voltage for an alternator?
13. Using Figure 13–8B, explain how an AC generator creates a sine wave.
14. What is the physical difference between a DC generator and an AC generator with a revolving armature?
15. What is the difference between an AC generator with a revolving armature and one with a revolving field?
16. What is the difference between a salient pole rotor and a turbo rotor?
17. Explain the relationship between speed of rotation, number of poles, and output frequency.
18. How do you increase or decrease the output power of an AC generator without changing its output frequency?
19. What are the requirements to be met before alternators can be paralleled?
20. Explain the use of a synchroscope.

PRACTICE PROBLEMS

1. Calculate the efficiency of a generator that has 15,000 watts of input energy and 3000 watts of internal power losses.
2. If a generator loses 3000 W of power in the generation process, and 35% is in friction, windage and iron loss, what is the copper loss in watts?
3. Find the frequency of a three-phase generator with 12 total poles, and a rotational speed of 1100 rpm.
4. Find the needed speed for a single-phase generator with 6 poles to operate at 40 Hz.
5. How many poles per phase are there in a three-phase generator if it operates at 60 Hz and 1800 rpm?
6. Why is an AC generator (alternator) that is to be parelleled supposed to run faster than the on-line alternator?
7. Name the various losses within a generator and what can be done to reduce the losses.
8. Changing the speed of an alternator changes the output voltage and also the _____.
9. By increasing the DC excitation to an alternator, the frequency does not change but the _____ of the waveform increases.
10. A voltage regulator on an alternator changes _____ _____ to change output voltage.

14

AC Motors

O U T L I N E

OVERVIEW

In this chapter you will be introduced to the AC electric motor and how it uses the power generated by alternators then converts the delivered electrical energy back into mechanical energy. As you have studied, the generator takes mechanical energy and converts it to electrical energy, which is transmitted to where it is needed; motors then convert the electrical energy back into mechanical energy. This process is very similar to the generator conversion in that there are three requirements for conversion: a magnetic field, a conductor, and a supply of current. The results of the interaction of these three elements will be motion.

For this chapter we will just explore what makes the motor turn. For detailed motor application, specific controls, and installation you should refer to a book on motors and controls.

OBJECTIVES

After completing this chapter, you should be able to:
- Explain motor action and the right-hand rule for motors
- Describe the process for creating a rotating magnetic field in a single-phase AC induction motor
- Define the term *slip* and relate it to induction motors
- Explain how to create a rotating magnetic field in a three-phase motor
- Use connection diagrams to connect three-phase motors in wye or delta patterns
- Calculate the line and phase voltages in a three-phase wye system
- Explain the basics of variable frequency drives (VFDs) and frequency effects of AC motors

MOTOR ACTION

The basis for an electric motor is that the relation of magnetic properties hold true as we apply current to a conductor and produce a magnetic field then bring two magnetic fields into close proximity. We start with the simple motor action of a conductor with electron flow when placed in a magnetic field. The conductor has a magnetic field developed by the current that reacts with the external field and the conductor tries to move out of the external field as in Figure 14–1. The direction of movement can be explained by the right-hand rule for motors. This right-hand rule uses the same finger denotations as the left-hand rule but changes the orientation as seen in Figure 14–2. As you note, the first finger on the right hand represents flux direction of the field.

FIGURE 14–1 Resultant motion with a current flow in a conductor in a magnetic field.

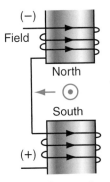
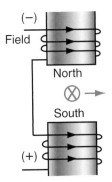
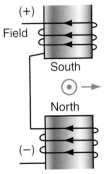

Original Connections Give Field Polarity as Shown. Armature Current as Shown Would Produce a Counterclockwise Rotation.

Armature Connection Changed to Give Opposite Direction Of Current in Armature While Maintaining Field Direction Results in Clockwise Rotation.

Reverse Field Polarity and Armature Polarity From Middle Diagram will Result in the Same Clockwise Direction of Rotation.

FIGURE 14–2 Right hand rule for motor direction when using electron flow theory.

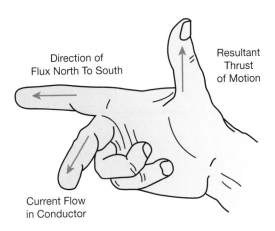

Direction of Flux North To South

Resultant Thrust of Motion

Current Flow in Conductor

(First = Flux direction as north to south). The center finger indicates current (Center = Current or electron flow in the conductor). The thumb indicates the direction of movement or thrust (Thumb = Thrust or direction of motion). Remember, the generator uses the thrust of the conductor through a magnetic field to produce current, and the motor uses the current and magnetic field to produce thrust. All electric motors work off this same principle.

SINGLE-PHASE AC MOTORS

All AC motors work because a rotating magnetic field can be produced on a stationary frame, or the stator, of the motor. We will discuss induction motors first as they are the most popular AC motor. The stator field can be created by having two sets of coils displaced in physical location around the rotor. If we start with a two-pole motor as in Figure 14–3, you will notice that there are two large poles created by the four coils with the electrical polarities as indicated at the coil ends. Each set of coils will have a separate AC sine wave supplied to it. To do this, the single sine wave needs to be split in to two separate waves. Figure 14–4 shows two waveforms split by 90 electrical degrees. The process of splitting the single phase into two waveforms is as follows: The motor windings (coils wound in the stator) are inductive. Some degree of phase shift occurs because the run winding is generally more inductive that the start winding, which has more turns of smaller conductor and is less inductive or more resistive. (Please note that series capacitors are also used in some motors to create a phase shift). The

FIGURE 14–3   Magnetic poles and electrical polarities for a two pole motor.

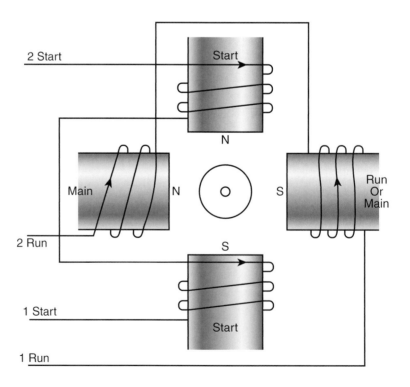

FIGURE 14-4 Phase shifted coil currents applied to motor coils.

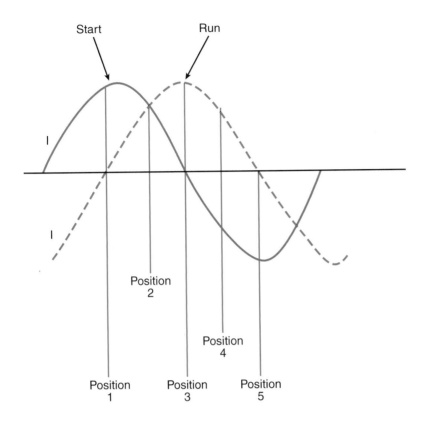

process of splitting the phase is different for different types of motors. If we apply the waveform marked "start" to the coils in Figure 14–3 marked start and the waveform marked "run" to the coils marked run we will create a two-pole electromagnet that rotates, or revolves, around the stator. If we use the position markings of Figure 14–4 and relate the current at this point to the coils in Figure 14–3, then we can assign the actual pole polarities using the left-hand rule for coils.

Explanation for Figure 14–4

Position 1: Peak current flow in the top and the bottom coils marked "start" with north on the inside edge of the top coil and south at the inside edge of the bottom coil. There is no motion yet as only the first two poles are established.

Position 2: Less current flows in the start coils as the sine wave diminishes and current increases in the side or "run" windings. With the electrical polarities at this point in time as shown, the inside face of the left pole is north and the inside face of the right pole is south. We still have two larger poles with the center of the magnetic pole shifted 45° counterclockwise.

Position 3: The top and bottom poles receive no current and produce no magnetic field. The side poles' "run" have maximum current and are producing maximum flux. The two poles have shifted their center another 45° to 90° counterclockwise from the starting point.

Position 4: The top and bottom poles now have current again but opposite to the original direction. Using the left-hand rule for coils, you can verify the magnetic pole faces have reversed from the original and the north pole has moved to the bottom inside edge. At the same time as this current is increasing, the side poles are weaker making the center of the north pole and the center of the south pole move another 45° counterclockwise.

Position 5: The side poles are again at zero current flow and are demagnetized. The start windings are strongest with the north pole at the bottom inside edge, another 45° movement. The two magnetic poles are rotating counterclockwise.

SYNCHRONOUS SPEED

Synchronous speed
The speed of the rotating magnetic field around the stator core, measured in revolutions per minute (rpm).

The speed that the rotating magnetic field rotates is dependant on the number of poles and the frequency of the waveform applied. The two poles rotate one full rotation around the stator core with one full waveform. With 60 Hz or 60 cycles in 1 second the poles would have 60 complete revolutions in 1 second. In 1 minute they would rotate 3600 times or 60 × 60. If fact, this is the **synchronous speed** of a two-pole motor connected to a 60-Hz supply. You can use the same formula used for generators with just an algebraic equivalent transposition. The formula for synchronous speed is:

$$S = \frac{(120 \times \text{frequency})}{\text{number of poles}}$$

where S is synchronous speed measured in revolutions per minute (rpm); 120 is a constant; converting pairs of poles to number of poles and 60 cycles per second to cycles per minute (same equation given in generation equations); Frequency is in hertz.

Number of poles means the total number of magnetic poles—not windings.

Example
$$S = \frac{120 \times 60}{2}$$ $$S = 3600 \text{ rpm}$$

Squirrel cage rotor
A rotating member of an induction motor with an iron core and conductors around the periphery. The conductors have induced voltage and resultant current flow as the rotating magnetic field spins on the stator. The stripped down rotor, without iron and only the conducting bars, resembles a squirrel cage exercise wheel.

INDUCTION MOTOR

A **squirrel cage rotor** as pictured in Figure 14–5, position 1, is made up of conductors that are mounted in the rotor and are short circuited at each end as depicted in the figure. There is no electrical connection between the rotor conductors and the stator conductors. However, there is current flow in the rotor conductors (bars) when the motor is operating.

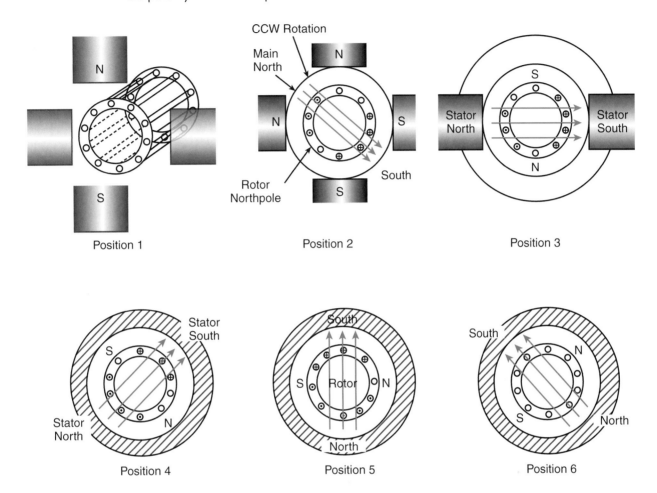

FIGURE 14-5 Magnetic poles are induced in a squirrel cage rotor based on the polarity of the stator poles.

Using Figure 14–5, you can understand how the voltage is induced in the rotor bars; how the current creates a magnetic field on the rotor; and why the rotor will follow the rotating field of the stator.

- Position 1: There is only a north and south pole on the stator and no voltage is induced into the rotor and no magnetic field is produced by the rotor. The rotor does not move.

- Position 2: As the side stator poles move 45° counterclockwise, the stator's magnetic field begins to cut through the rotor conductors. All three requirements for induction are present: (1) magnetic field, (2) conductor, and (3) relative motion. As the stator flux cuts through the rotor conductors it induces a voltage and resultant current in the circuit created by the conductors that are all electrically connected together at each end. Using the left-hand rule for conductors you can determine the direction of current flow as the magnetic field wraps around the conductors. The direction is shown on the rotor bars in the diagram. Now you can use the left-hand rule for a coil. As you see, the current comes toward you on the left side of the rotor and flows away from you on the right side of the rotor. This current will produce a

Slip

Slip refers to the difference in speed measured in rpm between the synchronous speed of the rotating magnetic field of the stator, and the slower speed of the induction rotor. It is often expressed as a percent using the formula:

$$\% \text{ slip} = \left(\frac{\text{sync speed} - \text{actual speed}}{\text{sync speed}} \right) \times 100.$$

FieldNote!

AC motors were invented by Nikola Tesla - a naturalized American citizen from Croatia who was an inventor, mechanical engineer, and electrical engineer. During the late 1800s and early 1900s he was a major force in the field of electricity and magnetism. Tesla did theoretical work on the concepts of AC generation and distribution, and on AC motors. His inventions enabled Westinghouse Electric and Manufacturing Company to capitalize on AC's advantages, especially an easier-to-maintain AC induction motor. Until his achievements were put into commercial use, motors used DC with its inherent problems; George Westinghouse was a leading promoter of AC, while Thomas Edison had been pushing for DC generation, transmission, and use in motors.

Tesla was involved in many groundbreaking new concepts, including radio. In 1943 - long after radio had first been in use - the United States Supreme Court declared him the inventor of radio. Many people have heard of a Tesla coil, which is used to produce high-voltage discharges. In 1893 - long before they came into everyday use - Tesla was showing people how fluorescent lamps worked. Tesla died in 1943, a forgotten and poor man.

north pole on the lower left of the rotor. This rotor north will be repelled by the stator north and the rotor will begin to move. The rotor north is also attracted by the stator south, increasing the magnetic action. The same actions of repulsion and attraction are occurring with the stator south and the rotor south.

- Position 3: As the stator poles continue to rotate, further cutting action takes place on the rotor conductors producing rotor poles as indicated by position 3 in Figure 14–5.

As the stator poles are moving the rotor conductors are cut, thereby creating an internal current flow in the rotor that continues to create rotor magnetic poles. The only way to induce a voltage into the rotor is if there is a cutting action created by the relative motion of the magnetic field. In other words, the rotor *cannot* spin as fast as the synchronous speed of the stator field. If the rotor does spin as fast, there is no cutting action, no induced voltage in the rotor, no current flow, and no magnetic poles develop on the rotor. With no magnetic poles the rotor has no magnetic attraction or repulsion to the stator poles. In order to induce voltage and resulting current in the rotor the rotor must **slip** in speed behind the speed of the rotating stator field.

Example

What is the percent slip of a motor that has a synchronous speed of 1800 rpm and an actual rotor speed of 1750 rpm?

$$\% \text{ slip} = \left(\frac{\text{sync speed} - \text{actual speed}}{\text{sync speed}} \right) \times 100$$

$$\% \text{ slip} = \left(\frac{1800 - 1750}{1800} \right) \times 100$$

$$\% \text{ slip} = \left(\frac{50}{1800} \right) \times 100$$

$$\% \text{ slip} = 2.78\%$$

MOTOR VARIATIONS

There are many different methods to produce the rotating magnetic field in the stator. Each method has its advantages and disadvantages such as cost, complexity, maintenance requirements, etc. as compared to better torque, better efficiency, and better operating characteristics. Some of the more common varieties of the single-phase induction motor are the split phase, the capacitor start, and the shaded pole. These motors can be connected for different voltages by arranging how the windings are connected to the line. A thorough study of each type of motor and how they are connected can be found in a motors reference book.

As you progressed from single phase to three phases in the generation of AC, you now can use the three phases generated to your advantage when connecting the three- phase motor. This type of motor is extremely popular in commercial or industrial applications where the phase supply of power is available.

THREE-PHASE MOTORS

As the name implies, the three-phase motors operate using all phases of a three-phase power system. Refer back to the three-phase generator section in Chapter 13 to review generation. The same three phases that are 120 electrical degrees apart are represented in Figure 14–6. You will use these three phases applied to the motor coil windings of a motor to create a three-phase (3∅) rotating magnetic field. See Figure 14–7 for a rendering of a three-phase, two-pole motor. This drawing represents a wye-connected motor demonstrating that the windings have one common point, just as the wye-connected generator you studied was connected. Motors can also be connected in a different pattern called a delta pattern. First we will discuss the wye pattern, as the transition from the generator to the motor is easier to follow.

TechTip!

Under normal circumstances, individual residential housing does not have three phases delivered to the home. There are three wires either "dropped" or run underground "lateral" to the home. However, this does not mean that there are three phases. Typically the three conductors represent two hot conductors and a neutral from a single-phase source. You will study this idea in more detail in Chapter 15. A three-phase system is composed of three "hot" conductors and usually a neutral which creates a three-phase, four-wire system.

FIGURE 14-6   Three phase sine wave produced by a three phase generator.

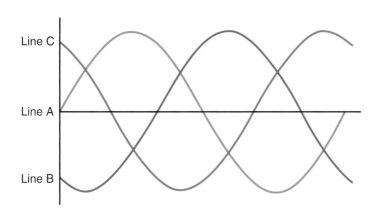

Line C

Line A

Line B

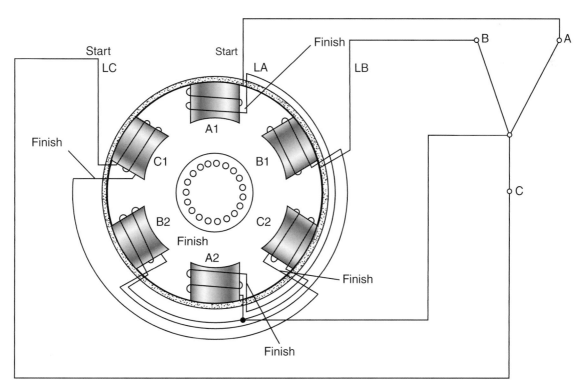

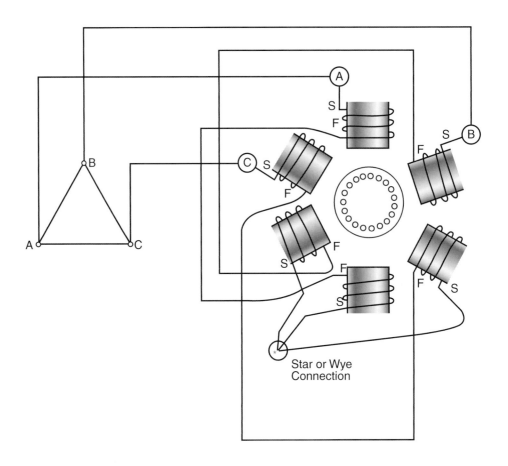

FIGURE 14-7   Three-phase, two-pole motor coil representation.

14.1 Rotating Magnetic Field of a Wye-Connected Motor

Refer to Figure 14–7 to determine how a magnetic field will rotate around the stator of a three-phase motor. Figure 14–8 will be used to explain how the applied voltage will create the magnetic poles of the stator. As you view the waveform, the waveform above the zero reference indicates current is flowing into the motor windings. When the waveform is below the reference, current is returning from the motor to the power source.

- Position 1 in Figure 14–8: Phase A has zero voltage and zero current. The top and bottom coils are not energized and have no magnetic poles. At the same instant in time phase C has approximately 70% of the maximum voltage applied and current is flowing into the coils for phase C. The voltage waveform also indicated that coil B is energized by 70% voltage, but current is flowing out of the motor toward the source. If you use the left-hand rule for coils and note the direction of current flow through the coils, you will find that the left coils for B and C both produce north poles, and the right two coils for B and C produce south poles. There are now two poles established—north on the left and south on the right.

- Position 2: At position 2 on the waveform graph, phase A has now increased to 50% voltage and current flows into the motor. Phase C has decreased to 50% voltage but the current is in the same direction.

FIGURE 14–8 Three phases are applied to motor coils and each position
indicates field motion.

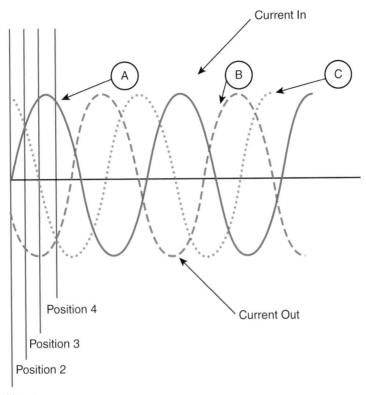

Phase B is stronger with 100% of the voltage and the current is
flowing to the source. As you can see, phase A has now established a
north pole at the bottom coil, phase B left coil has strong north poles,
and phase C left pole is weaker. The two poles'—north and south—
centers have moved counterclockwise.

- Position 3: Phase A has increased to 70% of maximum voltage, phase
  C is at zero, and phase B has decreased to 70% of maximum. This
  means that phase C has been demagnetized and produces no magnetic
  field. Phase B left pole is a little weaker north and phase A bottom coil
  is a little stronger north. The center of the north poles has moved
  counterclockwise again between phase A and B.

- Position 4: Phase A is at 100% voltage, phase B has decreased to 50%
  of the maximum, and phase C has reversed direction to 50% of the
  maximum and current is flowing opposite to the original direction. Of
  course this translates to the magnetic poles for C phase are opposite to
  what they were in position 1. Now the lower right C coil is at 50%
  strength and is north, phase A creates the strongest north at the bottom
  pole, and B phase has become a weaker north at the bottom left. The
  two poles have moved 90° counterclockwise from their original
  position at time 1, and are centered over the bottom and top poles.

As you notice the pole has rotated 90° through the stator positions and
when using phase A as a reference, the waveform has moved

TechTip!

When one of the three phases of a three-phase system is disconnected or fails, the system reverts to a single-phase system. The voltages between the phases are normally 120° apart and work in unison. As two conductors supply current to the load, the third conductor provides a path back to the generator. If you add all three currents together with the proper polarity (direction) the total must be zero. The phases are always equal and opposite. If one phase fails, the other two phases become the supply and return and are considered a single-phase sine wave. If this happens to a three-phase motor, the motor is said to be "single phasing" and the rotating magnetic field will not rotate as desired. Often if the motor is already running under light load, the motor will continue to spin. Under a heavy load, the motor will overheat and may stall. Typically a three-phase motor with a connected mechanical load will not start under a single-phase condition.

90 electrical degrees. Using this information you can deduce that the magnetic field will rotate through all 360 mechanical degrees as the phase completes the full 360° of electrical sine wave. Therefore for a three-phase motor, use the same formula for synchronous speed of the rotating magnetic field as was used in single phase. $S = \dfrac{120 \times \text{frequency}}{\text{number of poles}}$. The number of poles is counted as the number of poles per phase.

14.2 Three-Phase Delta Motor

Another popular three-phase motor connection for an induction motor is the delta connection. The method of connecting the coils wound in the motor stator is determined by the desired operating characteristics, the insulation ratings of the coil windings, the wire gauge of the coil conductors, as well as the manufacturer's preference. As you will see in Chapter 15, the voltage and current of the delivery system may be different than the voltage and current in the coils of the load. Figure 14–9 will help you determine why the connection of the coils is referred to as the delta connection, the coil pattern forms an equilateral triangle that resembles the Greek letter delta (Δ).

FIGURE 14–9   Schematic diagram of a dual voltage Delta connected motor coils.

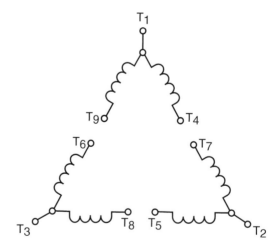

By viewing the delta pattern, each corner of the triangle represents one phase connection for the coils. In this example there is a phase A coil connected from T1 to T9, and another identical coil wound with the first coil connected from T6 to T3. In this dual-voltage motor, the coils can be connected in series as in Figure 14–10 to make them ready for twice the line voltage of the parallel connection shown in Figure 14–11.

FIGURE 14–10 Connection diagram for a high voltage Delta connected motor. Second diagram shows voltage drop across the coils of a dual-delta (series) connected motor configured for its high-voltage (480 volt) operation.

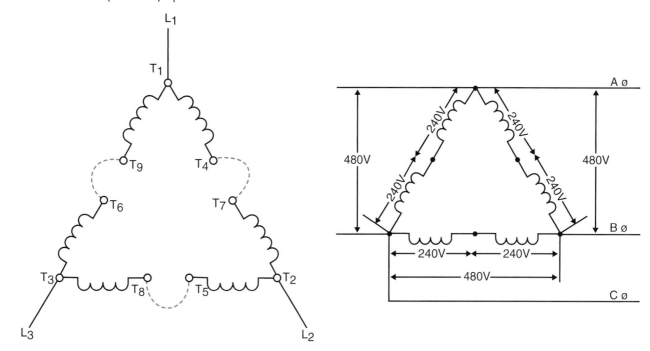

High-voltage Connection ------

FIGURE 14–11 Connection diagram for a low-voltage Delta connected motor. Second diagram shows voltage drop across the coils of a dual-delta (parallel) connected motor configured for its low-voltage (240 volt) operation.

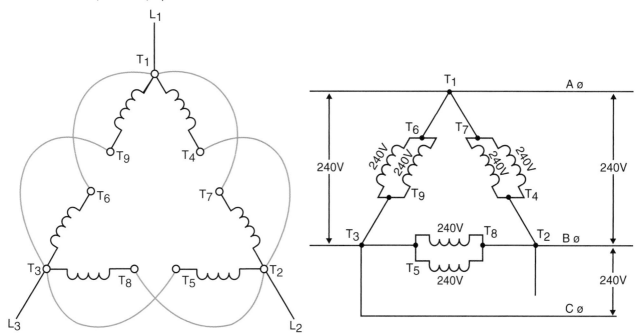

Low-Voltage Connection ———

THREE-PHASE INDUCTION ROTOR

Most three-phase motors are induction motors as opposed to conduction motors. The term *induction motor* simply means the rotors have voltage induced into them whereas *conduction motors* have voltage "conducted" or electrically connected to them. Three-phase induction motors use squirrel cage windings on the rotor just as single-phase motors use. Again, there needs to be relative motion of the magnetic field, meaning the rotor must slip behind synchronous speed of the stator field in order to have induction in the rotor. As the rotor creates magnetic poles through induction, the number of rotor poles will match the number of stator poles. In this basic theory book you have only seen examples of two-pole motors, but many more poles can be created. These are explained in further detail in motor books. It does not matter how many poles are on the stator or whether the stator is connected wye or delta, the rotor is the same basic squirrel cage.

GENERATOR EFFECT IN A MOTOR

As the rotor spins within the stator of a motor, it creates the magnetic poles on the rotor. These created magnetic poles act just like a generator with a rotating field. The rotor is a spinning magnetic field (moving magnetic field), and the magnetic field is cutting the conductors of the stator windings. This generator action in the motor has some interesting effects. The faster the rotor spins, the more generator action occurs. As the motor starts there is a large current draw to get the motor running. This is known as "starting" current or inrush current. Actually the high current at start is the result of the rotor not turning, or turning slowly, and therefore not generating back into the stator. You can determine that the generation is in opposition to the applied electromotive force (EMF) and is therefore considered counter EMF, or CEMF. As the motor comes up to speed, the rotor speeds up and the CEMF goes up, which causes the line current to the motor stator to go down—from the higher starting current. Of course the motor will still draw current from the line because the motor effect is greater than the generator effect.

Another recognizable effect of the generation effect of a motor is the noticeable effect of placing a larger mechanical load on the motor. This causes the motor to slow down, causing more slip which means more current is induced into the rotor and also means that the generation effect is lessened. A slower generator (spinning rotor) induces less CEMF into the stator and the current of the stator increases. You can verify that a motor working harder to produce more horsepower draws more current. Thus, you can see that every motor has some generator properties as well as every generator has some motor properties.

VARIABLE FREQUENCY DRIVES

Variable frequency drives, known as VFDs, are electronic controls used to control the speed of an AC motor. As was stated earlier the synchronous speed of a motor is determined by the number of poles per phase and the applied frequency. The actual speed is determined by the amount of slip required to maintain a speed lower than synchronous when using an induction motor. Because the number of poles is normally a design by the manufacturer and not normally adjusted in the field, the only other variable to speed control is to adjust the applied frequency. This is done by electronically altering the frequency to the motor.

The basic block diagram of the VFD is seen in Figure 14–12. The power input can be either single phase AC or more likely three phases, and is brought to the controller. A photo of a VFD is seen in Figure 14–13. The VFD has rectifier circuits that convert the AC to DC then filter it to smooth out the DC and set it to the desired voltage level. This is done in the DC control and power circuits. There are many different ways to change the DC back to a desired frequency of AC or "invert" the DC in the block labeled DC to AC inverter. The output of the VFD is a frequency and voltage needed to create the desired synchronous speed of the AC motor. Most VFDs are not appropriate for "capacitor start" or "split phase" single-phase motors. The reason is that these motors can be slowed to the point where the centrifugal switch can close and create serious operating problems with the motor. A centrifugal switch is found on single-phase motors and is used to disconnect the starting windings of a motor after it gets to the right speed. The switch opens the circuit to the starting windings, and keeps it open while the motor is running near rated speed. If the starting windings of a single-phase motor are allowed to stay in the energized circuit, or become reconnected, they will overheat and become damaged. More specific details are found in books about Motors.

FIGURE 14–12 Block diagram of operational components of electronic variable frequency drive.

FIGURE 14–13 Photo of typical VFD.

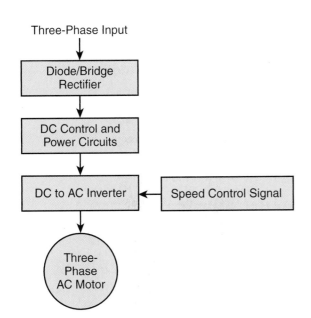

Two of the more common forms of inversion are known as current source input (CSI) and pulse-width modulation (PWM). The CSI-VFD has an output characterized by the waveforms in Figure 14–14. As you can see the voltage is not a clean sine wave and the resultant current is not a clean sine wave either. However, the motor still responds by creating the appropriate magnetic fields and moves the magnetic fields according to the formula for synchronous speed. The other popular form of inversion to AC is the PWM-VFD with the waveform as shown in Figure 14–15. By chopping the DC into small pulses and adjusting the width, or time of the pulse, the approximation of voltage appears closer to a sine wave. This is more apparent in the current sine wave as shown.

FIGURE 14–14 Representation of current source input (CSI) and current output waveforms.

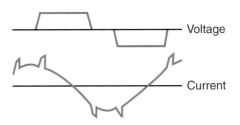

FIGURE 14–15 Representation pulse width modulation (PWM) and current output waveforms.

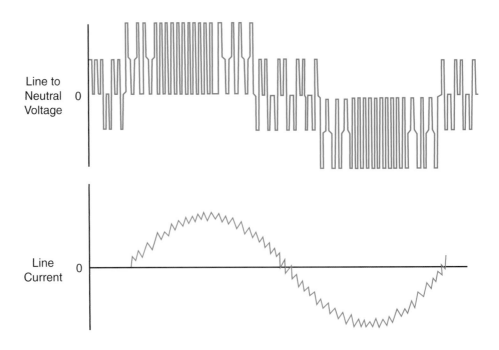

The VFDs are used in many control applications not only to control the running speed of the motor but to control other operating parameters. The starting speed and "ramp up" to full speed can be adjusted by controlling the frequency and allowing the frequency and speed to increase with time. The actual speed can be monitored and constant adjustments can be made to maintain exact speed under varying mechanical loads. The time required to stop can be controlled by a "ramp down" feature that will allow the motor to stop quickly, immediately, or control the deceleration to a specific time. VFDs are becoming more prevalent, more reliable, more versatile, less expensive, and have lower motor maintenance requirements. They have all but replaced the DC speed control systems that have been more accurate and more reliable for years.

SUMMARY

In this chapter you have learned that AC induction motors operate by inducing voltage and therefore current into a squirrel cage rotor. The method used to induce the rotor voltage is to create a rotating magnetic field at synchronous speed. The field moving through the rotor conductors creates a current and that current produces magnetic poles on the rotor. The rotor poles then are pushed and pulled by magnetism and the rotor follows the rotating stator field. Remember that there must be slip in the speed of the rotor as it cannot spin at synchronous speed.

Single-phase motors must create a phase shift in the single phase of the applied voltage. There are several ways to accomplish this, but the two resulting voltages are applied to the starting and running windings so that there is a rotation of the magnetic stator field. Depending on the particular motor style, the split could be between 30° and 80°. The speed of the rotating magnetic field, or synchronous speed, is computed by a single formula with the frequency and number of poles as the only variables. This formula is the same transposed formula used for generators.

The squirrel cage rotor was examined to see that there were no electrical connections to the rotor. The coupling to the rotor was only through magnetism even though there was current and voltage in the rotor. The percent slip was calculated to show that the actual speed of the rotor could vary. The rotor magnetically coupled to the stator will react to changing load and slow to create more magnetic field strength and more twisting effort known as torque. It was established that the rotor actually acted as a generator and generated a CEMF back into the stator windings.

Three-phase motors were explained to show the effects of applying three-phase voltage to a stator to produce a rotating magnetic field. The result of applying three phases to the motor creates a smoothly rotating magnetic field and allows the squirrel cage rotor to travel with the stator field. The same formulas for synchronous speed and slip apply equally to single-phase or three-phase motors.

Finally, an introduction to electronic control of motors through the use of VFDs was presented to explain how speed controls can be accomplished. A block diagram of the controller was used to show the sequence of the controller's functions. First the input waveform is rectified to create DC, and then the DC is set to the correct levels and inverted to create AC output. The variable frequency AC is directly connected to the motors to control the synchronous speed of the motor.

REVIEW QUESTIONS

1. Explain the right-hand rule for motors.
2. Equate the left-hand rule for generators to the right-hand rule for motors. How do they relate to each other?
3. Why does a single-phase AC waveform need to be split to operate as a single-phase motor?
4. How many actual magnetic pole windings are in a stator of a two-pole, single-phase motor?
5. How is the synchronous speed of an AC motor determined?
6. Explain why a squirrel cage rotor receives voltage and current without an electrical connection.
7. If a stator has four poles, how many rotor poles are established?
8. Is the rotor pole ahead, or behind, the stator pole of the same magnetic polarity in the direction of rotation?
9. The formula for percent slip is: _____.
10. Three-phase motors use this formula for synchronous speed: _____.
11. Show the two possible connections (low and high voltage) for the dual coils in Delta three-phase motors.
12. Why does a motor draw more line current if its mechanical load increases?
13. Why is the line current much larger for a motor starting than it is for running current?
14. Briefly describe the methods used for electronically controlling the speed of an AC motor.
15. How is PWM achieved in a VFD?

15

Transformers

OVERVIEW

As you have studied the effects and the characteristics of AC systems there has been one very consistent feature of AC circuits. The effects of induction are typically involved when we have a conductor, a magnetic field, and motion. You will see that a transformer is an excellent application for the AC waveform that is constantly moving as time progresses. The voltage is continually increasing or decreasing. You will now easily understand the operation of a transformer that has no moving parts.

This chapter will tie the theory of generation you have learned to the operation of motors and other devices that work on AC power. The transformer allows us to increase or decrease the voltage, or increase or decrease the current as we apply it to various situations, such as distribution systems and power supplies. The transformer allows us to "step up" the voltage and transmit power over long distances, then "step down" the voltage to deliver it to customers at appropriate levels.

This chapter will allow you to connect transformers in various circuit configurations to deliver the right voltage and current to a load. The transformer applications will allow you to form a basic understanding of an entire AC system from generation, to transmission, to distribution, and finally utilization.

OBJECTIVES

After completing this chapter, you should be able to:
- Identify parts of a transformer
- Explain how a transformer transfers power
- Calculate voltage, current, and power of a transfer based on formulas
- Describe losses in a transformer
- Identify types of coil and core arrangements
- Identify lead marking and additive or subtractive polarity transformers
- Connect transformer windings in basic circuit configurations

PRINCIPLES OF TRANSFORMATION

The basic premise of transformers is simple. The concept simply is to create a magnetic field with one conductor—coiled up—and allow that magnetic field to react with another conductor coiled up. The basic isolation transformer is shown in a schematic view in Figure 15–1. The conductor on the left receives voltage and current from a source. This conductor is typically wound around an iron core and is called the primary winding. The primary winding has all the properties of a coil studied so far. Namely it has resistance, exhibits inductive reactance, and has a magnetic field that surrounds the conductor and creates magnetic polarities. The magnetic polarities of the coil ends can be found at any instantaneous value of the AC wave by using the direction of current flow and the left-hand rule for coils. This magnetic field is constantly changing intensity as the AC current waveform constantly either increases or decreases. As the current changes, the magnetic field expands or contracts, and then changes direction. This AC magnetic field gives the needed component for induction; the component of movement.

FieldNote!

Transformers do not transform DC. The basic premise of a transformer is that it has no moving mechanical parts yet it induces voltage and therefore current into an adjacent coil of wire. The generator and a motor are examples of physical movement that allows the induction process to take place. A transformer only has the movement of an expanding and contracting magnetic field, powered by an AC current. DC is considered a steady value so the element of motion by the magnetic field is not achieved. You may occasionally see a transformer marked as "DC", a misnomer, but it is usually in part of an assembly working in conjunction with other components that create AC for the transformation or creates DC as part of a power supply.

For further reading, perform an internet search on "vibrators and step-up transformers" or "pulse transformers."

FIGURE 15–1 Schematic view of an isolation type transformer primary and secondary windings

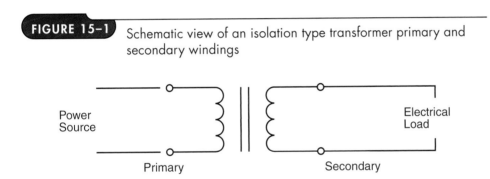

Power Source · Primary · Electrical Load · Secondary

Figure 15-2A is used to illustrate the action of primary current creating primary flux. Current flows into the negative terminal and out the positive terminal of the primary coil, as with other electrical loads connected to a supply. The primary flux, as verified by the left-hand rule for conductors, expands and is transferred to the secondary coils through the iron core material. As the primary flux moves through the secondary conductors, it induces a voltage with a direction of current flow as verified by the left-hand rule for conductors. If there is an electrical load connected to the secondary, secondary current will flow. Notice that the current in the secondary winding of the transformer flows from the positive terminal toward the negative terminal (see Figure 15-2A). The negative terminal of a supply is where current leaves the supply, flowing through the load and back to the positive terminal of the supply. Therefore, the secondary current leaves the negative terminal but flows positive to negative inside the transformer secondary. Refer to Figure 15-2B to see the effects of the secondary current flowing to the load. As secondary current flows, it creates its own magnetic flux. Using the left-hand rule for a coil verifies that the magnetic field created by the secondary current flow has a North Pole at the top, compared to the primary coil that has a North Pole at the bottom. The magnetic fields are in opposite directions. The effects of this interaction of magnetic fluxes can be determined when reviewing the waveforms in a later section titled "Characteristics of Transformer Operation."

FIGURE 15–2 (a) The primary current creates a flux that is transferred through the iron core to cut the conductors of the secondary winding. (b) The primary and secondary coil windings both create magnetic fields that are opposite in polarity to each other when current flows.

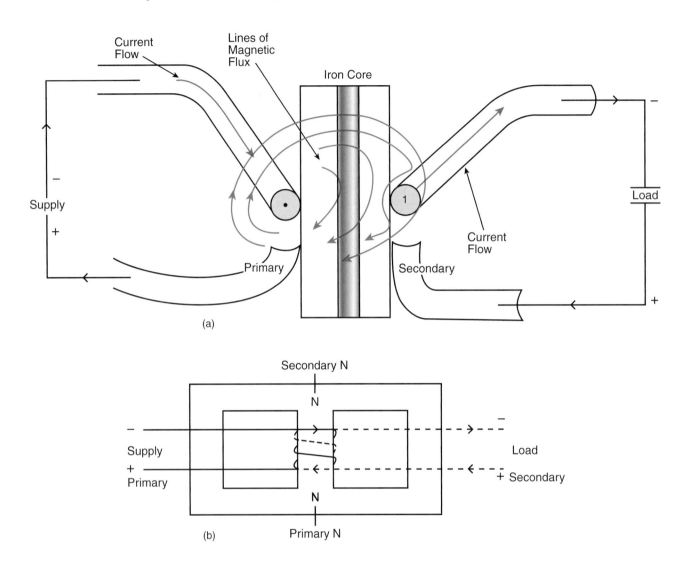

Referring back to Figure 15–1 you can see the electrical load is connected to the secondary. A load current will flow in the secondary circuit. The transformer has now transferred electrical energy from one circuit (primary) to another circuit (secondary) with no electrical connection. Only magnetic coupling is used to transfer the energy form one conductor to another.

VOLTAGE, CURRENT, AND POWER CHARACTERISTICS

Transformers are usually used to change the level of the voltage or current but are not designed to use electrical power or create electrical power. The main purpose of most transformers is to transfer energy while changing the voltage level of the circuit between primary and secondary. Using this concept of transferring energy without creating or using it allows us to understand the transformer action.

As the primary winding receives current from the source, it creates a flux based on the applied voltage and the number of turns of wire. The greater the voltage applied, the greater the current flow and, therefore, the greater the generated magnetic flux. This flux is distributed equally among the number of conductor turns to give a value of **volts per turn** as a comparative term. This tells us that that there is flux in the primary windings and the more windings—the lower the volts per turn—compared to a primary with few turns. Likewise, a large voltage applied to a fixed number of turns has a larger volts-per-turn ratio compared to the same number of turns and a lower voltage applied.

The volts per turn is an indicator of the amount of magnetic field produced by the primary. If there is good magnetic coupling between the primary and secondary coils, almost all the primary flux interacts with the secondary winding. In other words, the primary flux cuts through the secondary conductors. The secondary has the same amount of magnetic flux as the primary and it too has a comparative term of volts per turn. If this is the case then: (volts-per-turn primary) = (volts-per-turn secondary). By determining how many turns are used in the secondary and the volts-per-turn ratio you can determine the secondary voltage.

Another important concept is the reaction of the current. Assume a load is connected to the secondary and current is flowing. The maximum amount of secondary current that can flow without damaging the transformer will be based on the design and the fact that current is neither generated nor used in the transformation process. The voltage level between the primary and the secondary relates to the number of windings in the primary and secondary coils. The number of volts times the maximum amount of current in the primary yields one rating of the transformer measure in volt-amps. As you have studied, this is a measurement of apparent power. The apparent power indicates how much energy enters the primary windings. If essentially no power is deliberately consumed in the transformer, the same amount of apparent power is delivered to the secondary winding and consequently delivered to the load. Using this concept, if the voltage on the primary is higher than the voltage on the secondary then the current on the primary must be lower than the current on the secondary. The product of primary V × A must be the same as secondary V × A.

The power rating of the transformer is rated in volt-amps because the transformer is designed to transform voltage and current and has no influence on how that power is used. It could be used as watts, VARs, or VA. The windings are designed to transform the voltage level based on the number of windings, and the current is based on the circular mil size of the windings. Larger currents require larger conductors, and smaller currents only need smaller conductors. Therefore, the conductor size and the number of turns dictate the maximum current and voltage the transformer can handle without overheating or being destroyed.

Volts per turn
Volts per turn is used as a reference point for the amount of magnetic flux produced by a coil. (The more volts per turn the more flux.) It is not a specific unit of measure but a way to relate the primary turns and voltage to the secondary turns and voltage.

CALCULATIONS FOR PRIMARY AND SECONDARY

The calculations needed for transformer applications are basic algebra equations. The fact that power in = power out (VA in = VA out) out is considered the basis for all the other calculations. Although the statement is not entirely true, it is the assumption for almost all of the calculation. It is not entirely true because there are losses in the transformer in the form of poor magnetic coupling and heat in the core, created by eddy current, hysteresis, and copper losses.

TABLE 15–1   Efficiency of Single-Phase and Three-Phase Dry-Type Transformers

Single-Phase kVA	Efficiency	Three-Phase kVA	Efficiency
25	98.0%	30	97.5%
50	98.3%	75	98.0%
75	98.5%	150	98.3%
100	98.6%	300	98.6%
25	98.8%	750	98.8%

Note: Most distribution size transformers are rated in kVA. Instead of listing a transformer as 75,000 VA, it is listed as 75 kVA.

For most transformers, the amount lost in the transformer is inconsequential compared to the amount of VA transformed. Unless you are doing transformer efficiency calculations, you will generally ignore the losses as too small to worry about.

The formulas for voltage ratios, current ratios, and turns ratios are as follows:

$$\left(\frac{E_P}{E_S}\right) = \left(\frac{N_P}{N_S}\right) = \left(\frac{I_S}{I_P}\right)$$

where:
E_P is the voltage applied to the primary
E_S is the voltage developed on the secondary
N_P is the number of turns in the primary winding
N_S is the number of turns on the secondary winding
I_P is the current flowing in the primary winding
I_S is the current that could flow in the secondary winding

Example

If the primary winding has 100 turns and 100 volts applied and there are 200 turns on the secondary, what is the secondary voltage?

$$\frac{E_P}{E_S} = \frac{N_P}{N_S} \text{ is a ratio equation}$$

$$\frac{100}{E_S} = \frac{100}{200}$$

$$E_S = \frac{100 \times 200}{100}$$

$$E_S = 200 \text{ V}$$

Example

If the primary voltage is 240 V and the secondary voltage is 120 V, when there is 10 A in the secondary, what is the primary current?

$$\frac{E_P}{E_S} = \frac{I_S}{I_P}$$

$$\frac{240}{120} = \frac{10}{I_P}$$

$$I_P = \frac{120 \times 10}{240}$$

$$I_P = 5 \text{ A}$$

Example

If a 500 VA transformer has a 100 V primary and a 2:1 turns ratio, what is the voltage and current of the secondary?

$$P = E \times I$$

$$I_P = \frac{P}{E}$$

$$I_P = \frac{500}{100}$$

$$I_P = 5 \text{ A (current of the Primary windings)}$$

and

$$\frac{N_P}{N_S} = \frac{I_S}{I_P}$$

$$\frac{2}{1} = \frac{I_S}{5 \text{ A}}$$

$$I_S = \frac{5 \times 2}{1}$$

$$I_S = 10 \text{ A}$$

and

$$\frac{E_P}{E_S} = \frac{N_P}{N_S}$$

$$100 \frac{V}{X} = \frac{2}{1}$$

$$X = \frac{100}{2}$$

$$X = 50 \text{ V}$$

Verify:

$$100 \text{ V} \times 5 \text{ A primary} = 50 \text{ V} \times 10 \text{ A secondary}$$

The ratio of the currents is inversely proportional to the ratios for the voltage and the turns. This maintains the effect of changing voltages and resultant currents, yet maintains the VA rating from primary to secondary. The terms **step-up transformer** and **step-down transformer** refer to the transformation of voltage levels and not to current levels. A transformer that isolates the primary and secondary circuits is called an **isolation transformer.** This type of transformer simply separates electrical systems so that no electrical connection exists between the windings.

CHARACTERISTICS OF TRANSFORMER OPERATION

Transformers come in many different configurations. Refer to Figure 15–3 for a core-type transformer where the primary winding is on one leg of the core and the secondary is on another leg of the core. Notice in the diagram it appears as if the core is made of layers of iron sheets. In fact, the iron core is made of many laminations of thin iron pieces. The reason for laminating the iron is to reduce the losses due to eddy currents. Remember that eddy currents are internal electrical currents that flow through the iron as the magnetic fields are constantly changing and actually inducing current into the core. By laminating the core, the resistance to eddy currents is increased and the losses are reduced compared to a piece of solid iron. One of the other losses in the iron of the transformer is due to hysteresis. Hysteresis is the opposition to a changing magnetic field within a magnetized material. The molecules must realign each time the magnetic field changes, which is constantly, within the iron core of an energized transformer. The hysteresis losses are in the form of heat which is another watt loss in the transfer of power from primary to secondary. The use of low hysteresis steel (high-silicon content) reduces the losses due to hysteresis.

FieldNote!

In reality, the windings on a core-type transformer are wound on top of each other just as in any other type of transformer. Frequently the windings go most of the way around the core.

FIGURE 15–3 A representation of a core-type transformer.

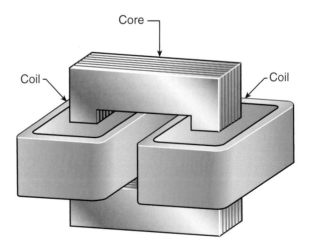

One of the most common types of transformer core is the shell type as shown in Figure 15–4. The lower-voltage winding, either primary or secondary, is wound first on the core. The reason for using a lower-voltage winding first is that the insulation on the **magnet wire** can be a lower rating and still not break down next to the grounded core. The higher voltage winding is wound on top of the other winding and there is nearly 100% flux linkage between the primary and the secondary winding. Figure 15–5 shows an H-type core. This core is a little more expensive to make but results in even less flux leakage and improves the efficiency of the transformer.

Magnet wire

Magnet wire has a coating of insulating varnish that provides the voltage rating for the wire. The wire is specifically designed to be used in coil winding of motors, generators, or coils. The wire fits into slots for easy winding and is used to produce a magnetic field.

FIGURE 15–4 A shell-type transformer with center windings.

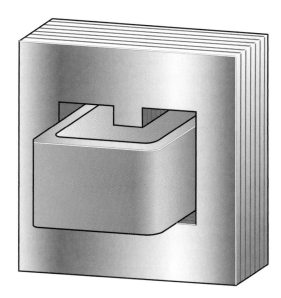

FIGURE 15–5 An H-type core transformer, often called a Type-H core transformer.

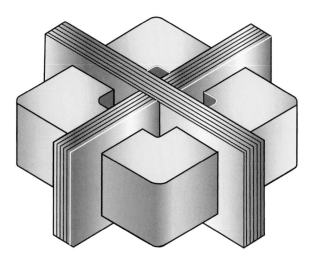

The basic function of each winding and the general location of the windings have been identified. The actual transfer of power through magnetic coupling can now be examined. Refer to Figure 15–6 to view the waveforms of voltage and current that occur in the primary and secondary windings. We make some assumptions for a perfect transformer even though we know the actual transformer waveforms are a little different. The primary voltage causes a primary current and a primary flux. We could see the effects of the flux in Figure 15–2A. With no electrical load connected to the secondary, the flux also produces a large CEMF in the primary winding and the current is impeded and, therefore, kept to a low value. This small current with no secondary current is the magnetizing current. As electrical load current flows in the secondary, the secondary current produces a magnetic flux opposite to the primary flux, because of the direction of the secondary current and the winding direction. The secondary flux is 180° out of phase with the primary flux, as seen in Figure 15–2B and Figure 15–6. The effect of the secondary flux is to reduce the effect that the primary flux has on the primary winding. The counter flux of the secondary reduces the CEMF of the primary winding by changing the magnetic effects on the coil. Now, with reduced CEMF on the primary circuit, more current flows in the primary. As we continue to draw more current on the secondary, the primary current also increases until the VA rating of the transformer design is exceeded. The power factor of the load will affect the transformer's actual waveform relationships between secondary and primary. The power factor of the load will not be transferred identically from the secondary to the primary.

FIGURE 15–6   Voltage, current and flux waveforms as they are derived in the primary and secondary circuits of a transformer.

Note: This illustration displays the E and I waveform relationships under no load. If an inductive load was added to the secondary, then the E and I waveforms would not be aligned or in-phase.

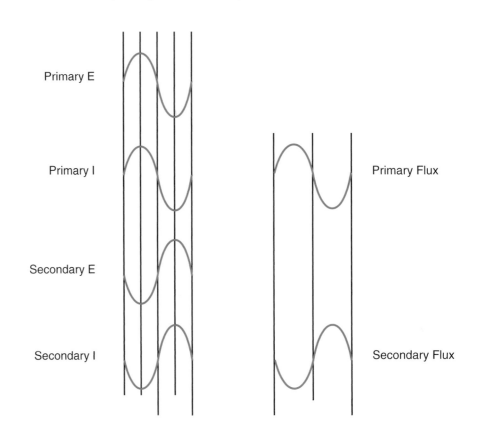

Compare the magnetic field sine wave of the primary to the magnetic field waveform of the secondary. The two fields are 180° out of phase (exactly opposite). When there is current flow in the secondary, the magnetic field of the secondary opposes, or cancels, some of the effect of the primary. With no current in the secondary, or no electrical load connected, the primary acts as a coil with a counter EMF (CEMF) produced by the magnetic self-induction, known as inductive reactance (X_L).

There is current in the primary but it is limited by the X_L to a small amount. This small primary current with no secondary current is referred to as **exciting current,** or **magnetizing current.** When a load is connected to the secondary current flows through the secondary coil and the magnetic field produced negates some of the effect of the primary flux. The reduced flux of the primary does not produce as much CEMF in the primary and the primary line current increases. The result is that as we draw more current from the secondary, the primary current also increases. This is a result of the interaction between the primary flux and the secondary flux.

Exciting current or magnetizing current
The relatively small current that is used to magnetize the core of a transformer but does not provide for the secondary load current is considered magnetizing current or exciting current. It is the primary current when no electrical load is connected to the transformer secondary.

TRANSFORMER IMPEDANCE

As mentioned, the impedance of the coils (the X_L and the R) in the transformer affects how the transformer works. One of the ratings for a transformer is the percent impedance rating. This rating is based on the full load ratings of the transformer and how much impedance is introduced to the power delivery system. The impedance is created in the transformer by these factors:

* type of core, which affects the percent of coupling;
* type of iron influencing permeability;
* wire size and type affecting wire resistance;
* number of turns of wire in the windings affecting the inductance.

The percent impedance determines how much current can pass through the transformer under the extreme conditions of a short-circuited secondary. If the transformer has a short circuit on the secondary, the impedance of the supply transformer will determine the available fault current at the transformer secondary.

To find the percent impedance of a transformer when not available on the nameplate, a test procedure is required. Figure 15–7 shows a single-phase transformer with a variable AC voltage source connected to the primary winding. The transformer secondary winding is shorted with an ammeter connected to measure the secondary current. The primary voltage is increased slowly from zero while watching the secondary amps. Increase the primary voltage until the secondary amperage is the full rated secondary current. This process of allowing current to flow in the secondary with only a small AC voltage in the primary will tell us how much the transformer impedes current flow to the secondary. To calculate percent impedance, use the formula:

$$\% \text{ impedance transformer} = \left(\frac{\text{AC voltage reading of primary source voltage}}{\text{rated primary voltage}} \right) \times 100$$

FIGURE 15–7   To test for transformer impedance, short circuit the secondary, then measure full secondary current with a small test voltage on the primary.

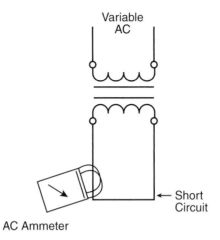

Variable AC

AC Ammeter

Short Circuit

Example

A 25 kVA single-phase transformer rated 480//240 volt is tested for percent impedance. The test source voltage reaches 24 VAC when the shorted secondary current reaches 104.17 A. What is the percent impedance?

Solution:

$$\% \ Z = \left(\frac{\text{test voltage}}{\text{rated voltage}} \right) \times 100$$

$$\% \ Z = \left(\frac{24 \ V}{480 \ V} \right) \times 100$$

$$\% \ Z = 5\%$$

The percent impedance is used to determine the compatibility of transformer paralleling, and for calculating the secondary fault current if a short occurs in the secondary. This fault current is used for calculations regarding arc flash hazards with power systems. Systems need design criteria for fuses and circuit breakers that can safely interrupt the fault currents available on the system. These ratings are found on circuit protection and other equipment as the Interrupting Current (IC) rating, or the fault current that the equipment must be able to withstand. The amount of energy available immediately in a short circuit accident is tremendous.

The formula for finding short circuit current, or fault current, when normal full voltage is applied to the primary is:

$$I_{(\text{short circuit})} = \frac{I_{(\text{full load secondary})} \times 100}{\% \ \text{impedance}}$$

Example

A 25 kVA single-phase transformer is rated for 480//240 V and a 5% impedance. How much fault current would the transformer have at the secondary terminals if a short circuit occurred?

Solution:

$$\frac{25000 \text{ VA}}{240 \text{ V}} = 104.17 \text{ A at full load}$$

$$\frac{104.17 \times 100}{5} = 52,085 \text{ A available short circuit current}$$

TRANSFORMER LEAD MARKING

Transformer leads (or bushings) are marked to provide the user with identification as the transformers are connected into different configurations of use. As you view the transformer, the higher voltage winding is marked with "H" designations and numbered H1 to H4, or other H markings. Look at Figure 15–8. The schematic view of the transformer shows the high voltage connection on the top side and starts with H1 on the top left. This is a consistent marking standard. The other side of the same coil windings is always H2.

FIGURE 15–8 Single phase transformer lead marking for single voltage and dual voltage windings.

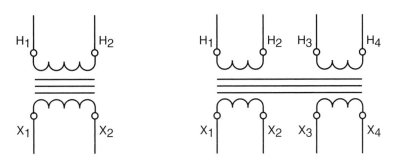

As you view the schematic the low voltage winding has "X" designations. The X designation varies as to where X1 starts—on the bottom left or the bottom right. The difference is a part of a marking agreement that categorizes the transformer into **subtractive polarity** or **additive polarity**. The standard identification for the low-voltage side is shown in Figure 15–9. The process has nothing to do with the transformer principles but only with the means of bringing the leads from the internal coils out to the external connection points.

Subtractive or additive polarity Polarity refers to the secondary lead markings when compared to the primary lead markings. With a subtractive polarity transformer, if the leads are brought out so that when H1 and X1 are connected together, a voltmeter between H2 and X2 will read the difference between the primary and secondary. An additive polarity transformer indicates that if H1 and X2 are connected together a voltmeter between H2 and X1 will read the sum of the two coil voltages.

FIGURE 15-9 Subtractive and additive polarity lead markings.

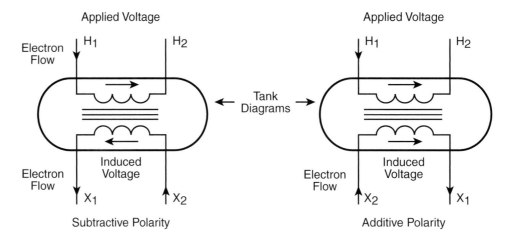

To determine the lead marking of a transformer where the secondary markings are not present or are not clear, use the process as shown in Figure 15–10. Use a small applied AC voltage to the high-voltage coil and place a jumper from H1 to the secondary lead that comes out of the transformer on

FIGURE 15-10 Tests for polarity and proper lead marking for additive or subtractive transformers.

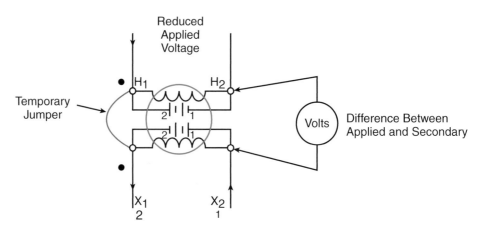

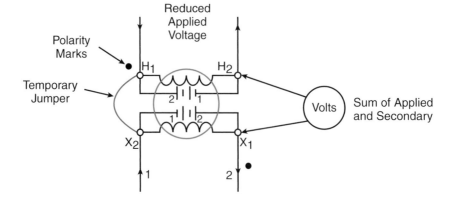

TechTip!

Figure 15–10 is a representation of the instantaneous voltages that occur in the transformer winding represented by batteries. The battery symbols are indications of the voltage values available at the same instant in time. The polarity marks are additional indications that tell the user where current enters the transformer (the marked H lead) and where current leaves the transformer secondary (the marked secondary lead). The subtractive polarity shows the two battery voltages opposing each other. The additive polarity shows the two battery voltages aiding each other. These permanent marks are helpful on instrument transformers, if the lead markings are destroyed.

the same side as H1. Now measure voltage from H2 to the other secondary coil lead. If the measured voltage is less than the applied voltage, then the transformer is wound as a subtractive polarity transformer and H1 is connected to X1. If the reading is more than the applied voltage the transformer is wound as an additive polarity and H1 is connected to X2. The small battery symbols shown on the transformer winding help explain the instantaneous voltage that is encountered during the measurement. It is important to note that the primary is a load to the supply and the polarities show electron flow from negative to positive. The secondary is a source of voltage and the internal voltage shows the current flow within the source going from positive to negative. Refer to the secondary electrical polarities and note that current is leaving X1 going to the load and returning to X2 from the load in both test connections. The battery polarity internal to the supply now makes sense and you can see why the two coil voltages oppose each other in the subtractive polarity and aid each other in the additive polarity.

Once the lead marking is established as correct, transformers can be connected in various patterns and the instantaneous polarities and the output waveforms will be synchronized.

SINGLE-PHASE TRANSFORMER CONNECTIONS

Single-phase transformers transform one sine wave of one voltage source at the input to another single sine wave of a new voltage source at the output. The coils that are wound on the core can be one single primary winding to one single secondary winding as already discussed. A variation of this process is to wind two separate coils for the primary and two coils for the secondary as illustrated in Figure 15–11. In this illustration the transformer is being used as a step-down transformer with each primary coil rated for 230 V and each secondary coil rated for 115 V. With this pattern the primary could be connected for up to 460 V by placing the two coils in series. Do this by connecting a jumper from H2 to H3 and applying 460 V to H1 and H4. The coils can be connected in parallel by connecting H1 to H3 (odd-numbered leads) together and connecting H2 to H4 (even-numbered leads) and applying 230 V across H1 and H2.

TechTip!

When electron flow enters H1 representing some point on the sine wave, electron flow will leave X1 at the exact same point in time. This fact is of little consequence when using a single isolation transformer. If the windings need to be connected in parallel or series, or if a three-phase circuit is constructed, the instantaneous current flow direction is critical. If the coils are connected together and expected to aid each other and the currents are flowing in opposite directions, then a short circuit results.

FIGURE 15–11   A dual-voltage primary and dual-voltage secondary for a single phase transformer.

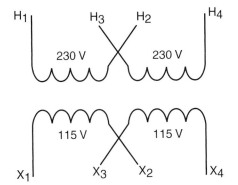

460-V Primary Connect H_2 to H_3

230-V Primary Connect H_1 to H_3, H_2 to H_4

230-V Secondary Connect X_2 to X_3

115-V Secondary Connect X_1 to X_3, X_2 to X_4

The secondary coils can be connected in series as well, X2 to X3 and the output voltage of 230 V is taken from X1 and X4. Parallel connections of the secondary coils (odd numbers connected together and then even numbers connected together) will produce 115 V output. So for this example, the voltage can be 460/230, 230/230, 230/115, 460/115. The primary is the first voltage and the secondary voltage after the double backslash. The rating of the transformer is rated with all windings connected to give the maximum kVA transformation.

If a single transformer does not have enough kVA to power the load, more transformers can be connected in parallel. Figure 15–12 illustrates a transformer with a single-voltage primary and a dual-voltage secondary. It is extremely important that the markings on the leads are correct for this type of connection. If the leads are not providing voltage in phase with each other, short circuits and damage can result. Notice that the H1 leads of both transformers are connected together as well as both H2 leads to place the primaries in parallel. Both secondaries are configured for the same secondary voltage and the X4 leads are connected in parallel as well as the X1 leads. The current splits at the primary energizing both primary windings and both secondaries add to the output current. The total kVA of the bank of two transformers is the sum of the two individual kVA ratings. The two ratings do not have to be identical, but the voltage input and output must be identical.

FIGURE 15–12 Two 10 kVA transformers in parallel will deliver 20 kVA to the load.

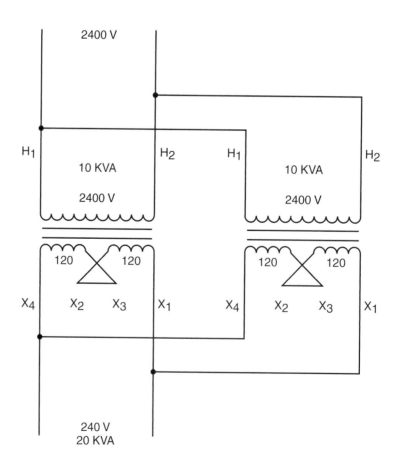

MULTI-COIL TRANSFORMERS

As previously discussed, transformers come in many different patterns for specific applications. One such style of transformer is a single-phase transformer used to step down distribution voltages for residential single-family dwellings. The transformer in Figure 15–13 is an example of this type of transformer, although the primary voltage can be much different from the standard normal rating depending on the distribution feeder. As you see in the illustration, the primary has many different tap points so that the volts per turn can be adjusted based on the actual voltage encountered in the field. In this case the primary voltage could be anywhere between 516 V to 444 V and the number of windings on the primary can also be adjusted from maximum to minimum. The effect is to create the correct volts per turn on the primary to yield the proper flux and produce 240 V across the two secondary coils.

FIGURE 15–13 A transformer with multiple taps on the primary to adjust for various supply voltages.

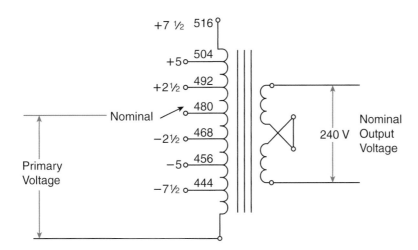

Note: This transformer has 6 taps with three $-2\frac{1}{2}\%$ taps for voltage below nominal and three $-2\frac{1}{2}\%$ taps for voltage above nominal. These terms are used in manufacturer's descriptions of their transformers.

A very standard secondary voltage system is created as 240 V across two coils with a connection point in the middle that is grounded. This establishes a three-wire, single-phase supply rated at 120 or 240 V with a nominal 115 to 230 V. Figure 15–14 shows another transformer with loads connected to the secondary three-wire system. Some loads operate at 120 V as in L1 to neutral or L2 to neutral and some loads are connected directly from L1 to L2 for 240 V. If you were to view the 240 V sine wave it would appear as one large sine wave. The waveforms from L1 or L2 to neutral also appear as a single sine wave at one-half the amplitude of the L1 to L2 waveform.

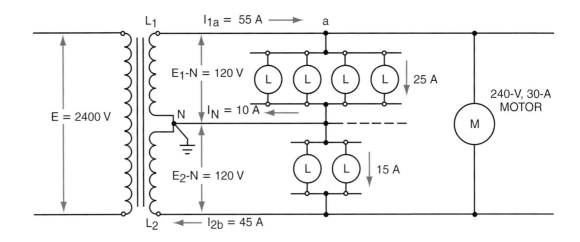

FIGURE 15-14 A single phase 3 wire system with loads at 120 V to neutral and 240 V line to line.

THREE-PHASE TRANSFORMER CONNECTIONS

You now know how single-phase transformers operate using mutual induction and the designed ratio between the primary and secondary windings. Now you will perform the same transformer action but with three-phase systems. Figure 15–15 is a pictorial view of three single-phase transformers connected to provide all three phases for a secondary load of a three-phase motor and several single phase loads. The system depicted in Figure 15–15 is called a wye-wye (Y-Y) because the schematic view of the coils appears as in Figure 15–16 and resembles the capital letter Y. As can be noted in the pattern, each transformer transforms one phase. The voltage on each coil is 2400 V on the primary. The line-to-line voltage is actually 1.732 ($\sqrt{3}$) times the phase voltage of 2400 V. This is the nominal line voltage of 4160 V but only 2400 V is across each coil from H1 to H2. This factor of 1.732 is a constant used in three-phase systems because adding voltages arithmetically is not possible as the voltages are 120° out of phase.

Figure 15–16 also shows the secondary connected in a wye pattern with the center point of the wye grounded to create a neutral point reference to ground. This connection forms a three-phase, four-wire wye system. Each transformer steps each phase down from 2400 to 120 V or at a 20:1 ratio. The secondary coils are delivering 120 V each and an individual sine wave is 120° apart from the other sine waves. We can again add the phases that are 120° apart by vector addition. The same factor of $\sqrt{3}$ or the agreed upon factor of 1.732 is used to determine the line-to-line voltage: 120 V × 1.732 = 208 V. Now you can see how 120 V single-phase loads are connected from line to neutral or across the coils of the secondary and the motor is connected for 208 V and it uses all three phases to operate as an induction motor.

FIGURE 15–15 Y-Y transformer pattern for three phase and single phase loads.

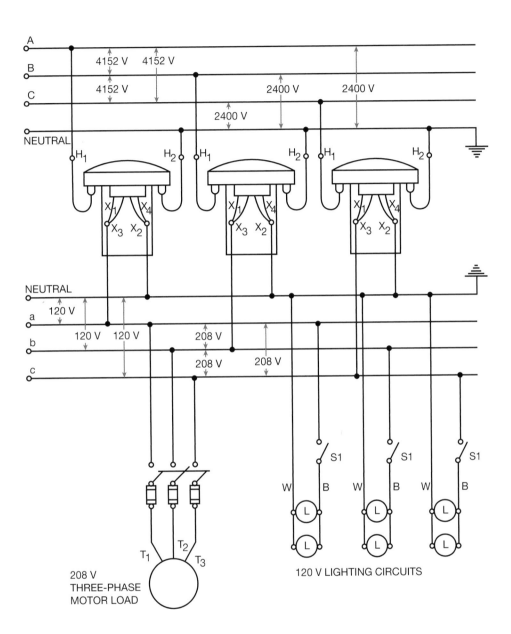

FIGURE 15–16 Y-Y schematic pattern with example voltages.

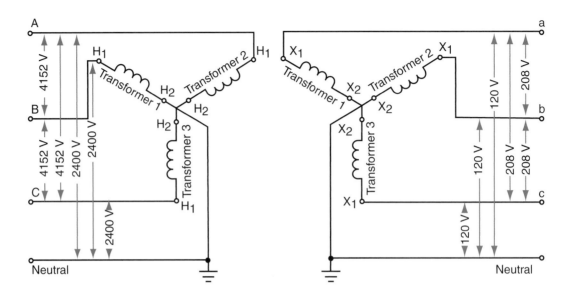

The other major connection pattern is the delta connection. Refer to Figure 15–17 for a wiring diagram view of the connection as a three-wire delta primary and a four-wire delta secondary. Figure 15–18 depicts schematically how the three individual phases are connected. With this pattern in the secondary we have a lot of three-phase power and only one phase supplying the single phase of 120 to 240 V. This pattern, known as the four-wire delta, has another voltage that is peculiar to this arrangement. The voltage from the ground to the top phase connection is 1.732 × the line to neutral voltage. For this example the "phase with the higher voltage to ground" known as the "high leg" or "wild leg" is 1.732 × 120 V = 208 V. Care must be taken not to get this leg mixed up with the other two when connecting 120 V loads.

The line-to-line voltage in a delta pattern is the same as the phase voltage, unlike the arrangement in a wye system. The VA rating of these three-phase patterns is three times the single phase kVAs if all the single-phase transformers are the same.

There are many connections and transformer calculations to create the desired voltages, currents, and kVA required by the load. These two patterns are examples of how the voltage can be transformed to yield various voltages and supply various loads. Books on transformers will be used to study transformers in more detail as your education continues.

FIGURE 15–17 Δ-Δ (delta-delta) connection pattern for three single phase transformers is a three phase arrangement.

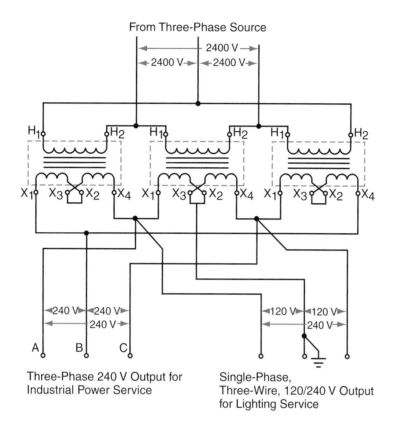

From Three-Phase Source

Three-Phase 240 V Output for Industrial Power Service

Single-Phase, Three-Wire, 120/240 V Output for Lighting Service

FIGURE 15–18 A delta-delta pattern with a three phase four wire secondary providing three phase and single phase power.

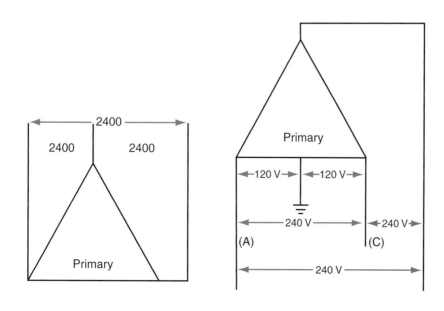

SUMMARY

This chapter is a primer on transformation of AC voltages and currents. This is by no means the complete coverage of transformer types, styles, or connection patterns. You have studied the basic concepts of transformer action that uses the three conditions for induction with a moving magnetic field providing the motion. A discussion focused on how voltage, current, and power are transferred through the transformer without an electrical connection. The supporting calculations for basic transformer applications have been provided.

Coil characteristics and transformer lead markings that allow you to connect transformers in standard connections were also discussed. Slightly more complex multi-coil transformers were presented to allow you to see that many variations and applications are used to provide the correct current, voltage, and power to the load being served. Single-phase transformers and popular three-phase connections allow you to see how common voltages you need for electrical loads such as motors can be delivered from AC generation systems. Something very important to the installer is the voltage taps designed into the transformer. The transformer is designed to operate at nominal primary voltage, such as 480 volts. Because of line resistance and other plant loads, however, the actual primary voltage may be 468 volts. Voltage taps allow the installer to accommodate this problem.

REVIEW QUESTIONS

1. Describe what is meant by primary and secondary windings of a transformer.
2. Explain what is meant by a step-up transformer.
3. Why would an isolation transformer be used?
4. If the secondary of the transformer has one-half the turns of the primary, what is the relation to the transformer voltages?
5. A transformer has 10 amps in the primary. The turns ratio is 1:4. What is the secondary current under full load?
6. A transformer has 120 V and 10 A on the secondary. What is the transformer primary VA rating?
7. Explain why transformers are not rated in watts.
8. Explain why the current ratio is inverse to the voltage ratio.

9. Why does the primary circuit have current even when there is no load in the secondary?
10. Why is the polarity of the transformer important?
11. When connecting three-phase transformers in a wye pattern and the coil voltage is 120 V, what is the line-to-line voltage?
12. Using Figure 15–19, show how to connect a transformer for 480 V primary and 120 V secondary.
13. Explain what is meant by a "high leg" on a transformer secondary.
14. What is the total kVA rating of two 50 kVA transformers connected in parallel with each other?
15. Why are transformer cores made of laminated steel?

FIGURE 15–19   Connections for dual voltage primary and secondary connections for Question 12.

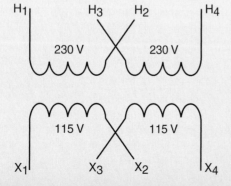

460-V Primary Connect H_2 to H_3

230-V Primary Connect H_1 to H_3, H_2 to H_4

230-V Secondary Connect X_2 to X_3

115-V Secondary Connect X_1 to X_3, X_2 to X_4

Mathematics: Using Vectors Effectively

OVERVIEW

Circuit analysis and calculations with resistors can be easily performed using simple arithmetic, addition, subtraction, multiplication, and division. The only concern is to be certain that you pay attention to the sign of the calculation, that is, negative or positive. This is true whether the circuit being analyzed is AC or DC.

Mixing AC resistors, inductors, and capacitors into your calculations complicates the issue. Analysis of such circuits requires a more sophisticated approach to the mathematics, namely, the use of vectors.

In this appendix, we will review how to solve systems vectors by adding, subtracting, multiplying, and dividing them. A proper understanding of vectors is necessary for you to understand the behavior of the variables in AC circuits that contain resistance, inductance, and capacitance.

Look at Figure A–1. The **reactance** (3 Ω) and **resistance** (4 Ω) in a circuit are at a 90° angle to each other and are acting on a common point. Vector analysis provides the means for calculating the combined or net effect of these two forces. In this example, the resulting combination is equal to a magnitude of 5 Ω and an angle of 36.9° from the horizontal axis.

Reactance
The amount of opposition to current flow exhibited by a magnetic field in an inductor or an electrostatic field in a capacitor.

Resistance
The amount of opposition to current flow through a material caused by so-called frictional effects.

FIGURE A–1   Vector result of a reactance and a resistance in the same circuit.

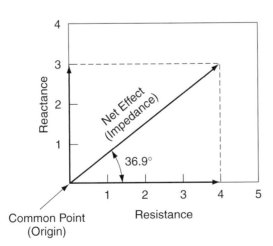

Vectors are also used to show the phase relationships between voltage and current in AC circuits; therefore, you must rely heavily on vector mathematics to solve circuits having inductive and capacitive components.

This appendix also covers information on solving for the resultant vector value for any combination of multiple vectors using several different methods to solve those problems.

SCALARS, VECTORS, AND PHASORS

A.1 Scalars

Scalar
A number with magnitude only. For example, 5 miles per hour is a scalar.

Scalars are numbers used to represent magnitude only and do not take direction into consideration. Assume that you are lost in the woods. As you walk down the path, you come upon a sign like that shown in Figure A–2. Does this sign help? It tells you that it is 3 miles to the highway, but in which direction? This sign would represent a scalar because it contains only a magnitude. If the sign were firmly planted in the ground and had an arrow painted on it pointing toward the highway, it would become a vector because it contains both the magnitude and the direction. Figure A–3 represents a vector having both magnitude (3 miles) and direction (direction of arrow).

FIGURE A–2 Scalar path sign.

FIGURE A–3 N-type and P-type materials.

A.2 Vectors

Vector
A number with both magnitude and direction. For example, 5 miles per hour going due east is a vector.

A **vector** is a symbol that indicates both magnitude and direction. A vector tells not only *how much* but also *in what direction*. A vector is represented graphically by an arrow. The length of the arrow represents the magnitude. The tip of the arrow represents the direction of the vector and is identified by its angle of rotation from 0°. For example, the direction of the vector in Figure A–1 is 36.9° counterclockwise from the horizontal axis.

A.3 Phasors

Phasor
A vector that rotates. Phasors are used to describe voltages, currents, and other such quantities in electrical systems.

A **phasor** is a vector that rotates. A voltage and/or a current in an AC circuit will be represented by a phasor that rotates at the frequency of the AC waveform. Although phasors are somewhat more powerful and used differently than vectors, manipulation of them in the circuit analysis that you will be working with is identical to vectors. The remainder of this chapter and text refers to all electrical quantities as vectors even though some of them are, strictly speaking, phasors.

A.4 Vector Referencing

As stated previously, a vector is drawn as a straight line with an arrow placed at one end. The arrow represents the direction of the vector, and the line length represents the magnitude of the vector. The vector can represent any quantity, such as inches, miles, volts, amps, ohms, or power.

The zero reference (0°) is a horizontal line to the right. The direction or angle of other vectors is positive when measured in a counterclockwise direction (see Figure A–4). In Figure A–4, a vector with a magnitude of 4 is at angle 0°. The second vector is drawn with a magnitude of 3 at an angle of 45° from the first vector. Now the third vector is referenced from the first vector at an angle of 90°. It is important to note that the third vector is referenced from the 0° line and not the previous vector at 45°.

FIGURE A–4   Vector counterclockwise rotation.

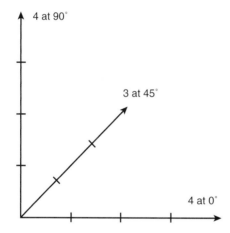

Notice that the three vectors could also be referenced in a clockwise direction using the negative values. The first vector would still be 4 at 0°, the second would be 3 at -315°, and the third would be 4 at -270°.

A.5 Rectangular and Polar Representation

So far in this chapter, all the vectors have been represented in what is called **polar form**: a magnitude and an angle. Vectors used in AC circuit analysis can also be referenced by expressing them as the vector sum of two other vectors that are at 90° from each other. This is called **rectangular form**.

For example, the vector shown in Figure A–1 can be expressed in polar form as 5 at an angle of 36.9°. This is usually written as 5 ∠ 36.9°. The vector in Figure A–1 can also be represented as the sum of one vector at 0° plus another vector at 90°. In this case, the 0° vector is 4 Ω long, and the 90° vector is 3 Ω long. The entire expression is usually given as $4 + i3$ or $4 + j3$, where $=i$ or $+j$ means that the second vector goes up vertically or 90°.

Polar form
Representation of a vector as a magnitude and an angle.

Rectangular form
Representing a vector as the sum of two other vectors that are at right angles to each other.

RIGHT TRIANGLES AND TRIGONOMETRY

To manipulate vectors effectively, the electrician must be familiar with the characteristics and behavior of right triangles and the relationships of the sides of a right triangle. The next two sections describe these important concepts and will serve as a refresher for your (probably) long-forgotten trigonometry studies.

A.6 Right Triangles

Hypotenuse
The side of a right triangle that is opposite the right angle.

Look at Figure A–5. A right triangle is a triangle that has a 90° angle. The **hypotenuse** is the longest side and is always the side that is opposite the 90° angle. Pythagoras of Samos figured out that the sum of the square areas bordered by each of the two sides of the right triangle would be equal to the square area bordered by the hypotenuse, or longest side. This is stated mathematically as $C^2 = A^2 = B^2$.

Where:

C = the length of the hypotenuse

A = the length of one side

B = the length of the other side.

FIGURE A–5   Right triangle with the sides squared.

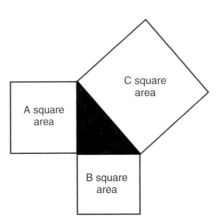

The previous equation is called the Pythagorean theorem in honor of its discoverer. Using the Pythagorean theorem, the hypotenuse in Figure A–5 can be found as:

$$8^2 + 6^2 = C^2$$

$$64 + 36 = C^2 = 100$$

$$C = 10$$

Example

Look at Figure A–6. Which side is the hypotenuse? What is the length of the hypotenuse if two of the sides have a value of 3 and 6?

Solution:

$$C^2 = A^2 + B^2$$

$$C^2 = (6)^2 + (3)^2$$

$$C^2 = 36 + 9$$

$$C^2 = 45$$

$$C = \sqrt{45} = 6.71$$

FIGURE A–6 Example of Pythagorean theorem.

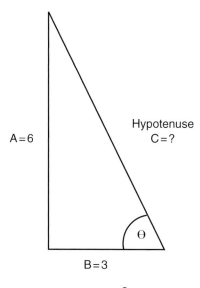

A = 6

Hypotenuse
C = ?

θ

B = 3

$$A^2 + B^2 = C^2$$

A.7 Sines, Cosines, and Tangents

Since the sides of a right triangle have a mathematical relationship to each other (the Pythagorean theorem), it should come as no surprise that the ratios of the various sides of the right triangle also have very specific values.

Figure A–7 is a right triangle with the various sides and angles labeled. Each angle has a side opposite and a side adjacent to it. Table A–1 shows the relationships of each angle to each of the sides. Use Figure A–7 and Table A–1 to understand the definitions given in Table A–2.

FIGURE A–7

Sample triangle for sine, cosine, and tangent.

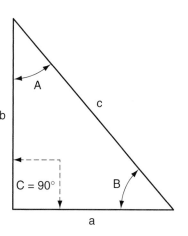

b

A

c

C = 90°

B

a

TABLE A–1	Angles and Their Related Sides (see Figure A–7)	
Angle	**Side Opposite**	**Side Adjacent**
A	a	b
B	b	a

TABLE A–2 Table of Trigonometric Definitions (see Figure A–7)

Angle	Sine	Cosine	Tangent
A	$\sin(A) = \dfrac{\text{Opposite}}{\text{Hypotenuse}} = \dfrac{a}{c}$	$\cos(A) = \dfrac{\text{Adjacent}}{\text{Hypotenuse}} = \dfrac{b}{c}$	$\tan(A) = \dfrac{\text{Opposite}}{\text{Adjacent}} = \dfrac{a}{b}$
B	$\sin(B) = \dfrac{\text{Opposite}}{\text{Hypotenuse}} = \dfrac{b}{c}$	$\cos(B) = \dfrac{\text{Adjacent}}{\text{Hypotenuse}} = \dfrac{a}{c}$	$\tan(B) = \dfrac{\text{Opposite}}{\text{Adjacent}} = \dfrac{b}{a}$

Note: Sine is abbreviated as *sin*, cosine is *cos*, and tangent is *tan*.

Several memory tricks are often used to remember this relationship. **Os**car **H**ad **A** **H**eap **O**f **A**pples or **O**h **H**eck **A**nother **H**our **O**f **A**gony are two of the many. They work as shown in the following equations:

$$\frac{\text{Opposite (Oscar)}}{\text{Hypotenuse (Had)}} = \sin$$

$$\frac{\text{Adjacent (A)}}{\text{Hypotenuse (Heap)}} = \cos$$

$$\frac{\text{Opposite (Of)}}{\text{Adjacent (Apples)}} = \tan$$

Some believe that remembering these clever memory devices is more difficult than remembering the original definitions. You should use whatever method is easiest for you.

After the sine, cosine, or tangent of the angle is known, the angle itself can be found using the trigonometric functions on a scientific calculator or by the trigonometric tables.

Example

The hypotenuse of the triangle in Figure A–8 is 14 and side *a* is 9. How many degrees are in ∠ A?

Solution:
Since the lengths of the hypotenuse and the opposite side are known, the sine function can be used:

$$\sin(A) = \frac{\text{Opposite}}{\text{Hypotenuse}} = \frac{9}{14} = 0.643$$

But 0.643 is not the angle; it is the sine of the angle.

You must use a scientific calculator and the inverse sine function (sometimes called the arcsine) to determine the angle. The way you determine the inverse sine depends on the type of calculator you are using. Either enter 0.643 and press the arc or inv sine button or enter arc sine (0.643) and press the equals (=) sign. The answer is arcsine (0.643) = 40°.

The same process is used when utilizing the cosine function or the tangent function.

Example

Using the same triangle in Figure A–8, find the number of degrees in B.

Solution:

$$\cos(B) = \frac{\text{Adjacent}}{\text{Hypotenuse}} = \frac{9}{14} = 0.643$$

The scientific calculator shows that the arccosine of 0.643 is 50°.

Note in this problem that the opposite side became the adjacent side to the hypotenuse because the reference angle changed. For this problem, side a is the base of the triangle. Also note that the decimal (0.643) remained the same because of the ratio (9/14), but the angle changed because of the use of the cos function instead of the sin function.

FIGURE A–8

Triangle for Example.

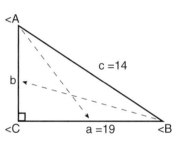

One last point should be noted. All the angles in a right triangle must total 180°. This means that the right angle ($\angle C = 90°$) + $\angle A$ + $\angle B$ = 180°. If you knew from the first problem that $\angle A$ was 40°, then you could calculate the remaining angle by simple addition and subtraction.

Since 90 + 40 = 130, and 180 − 130 = 50, then $\angle B = 50°$.

VECTOR ADDITION AND SUBTRACTION

Vectors can be added and subtracted in a variety of arithmetic and geometric ways. The following sections describe those methods and give examples of how each may be employed.

A.8 Vectors in the Same Direction

Since vectors can be used to represent quantities such as volts, amps, ohms, and power, they can be added, subtracted, multiplied, and divided. There are several methods to perform vector addition. Regardless of the method used, they must be added with a combination of geometric and algebraic addition. This is called vector addition. (Note that vector addition is quite easy on most scientific calculators; however, you should understand the fundamentals.)

One method is to connect one vector to the endpoint of the other one. This works easiest when the vectors are in the same direction (see Figure A–9). In this figure, three vectors with the same angle (0°) are being added. The total sum is 7 + 5 + 3 = 15.

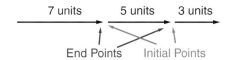

FIGURE A-9   Simple vector addition.

Now look at Figure A–10 and consider having two batteries connected in series the way they are in a flashlight. The bulb in the flashlight is designed to operate on 3 volts. Since the standard flashlight cell is only 1.5 volts, two batteries must be added together to make 3 volts. In vector terms, it would look like one vector of 1.5 volts plus another vector of 1.5 volts (see Figure A–11).

FIGURE A-10   A flashlight with two batteries.

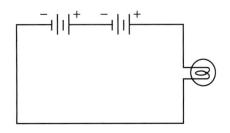

FIGURE A-11   Two 1.5-volt vectors added together to get a total of 3 volts.

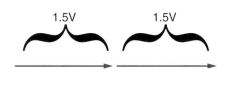

A.9 Vectors in Opposite Directions

To add the vectors that are in opposite directions (180° apart), subtract the magnitude of the smaller from the magnitude of the larger. The result is a vector in the same direction as the vector with the larger magnitude. Assume that the batteries in Figure A–10 are 3 volts and 5 volts, respectively. Somehow one of the batteries was placed in the flashlight backward. The voltages would oppose each other. This means that 3 volts of the 5-volt battery would try to overcome the 3 volts of the other battery. The result would be a vector with a magnitude of 2 volts in the same direction as the 5-volt battery (see Figure A–12).

FIGURE A-12   Adding vectors with opposite directions (180°).

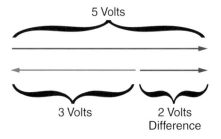

In numerical form, this equation would look like adding a positive number to a negative number:

$$+5 +(-3) = +2$$

Since adding a negative is the same as subtracting, the previous equation reduces to:

$$5 - 3 = 2$$

A.10 Vectors in Different Directions

Vectors having directions other than 0° and 180° from each other can also be added. An example is shown in Figure A–13. A vector with a magnitude of 4 and a direction of 20° and a vector with a magnitude of 3 and a direction of 45° are added together. The addition is made by connecting the starting point of the second vector to the endpoint of the first vector. The direction of the vector, stated as the number of degrees of the angle, is always referenced to the x-axis which will always be 0°. The resultant is drawn from the starting point of the first vector to the endpoint of the second. This is called the triangular method. It is possible to add more than two vectors in this fashion.

FIGURE A–13   Adding vectors with different directions.

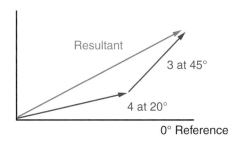

Resultant

3 at 45°

4 at 20°

0° Reference

A.11 Parallelogram Method

The parallelogram method can be used to find the resultant of two vectors that start at the same point instead of connecting the starting point of the second vector to the endpoint of the first vector. A parallelogram is a four-sided figure whose opposite sides are parallel to each other. A rectangle is a parallelogram with 90° angles.

For example, consider the vectors 10∠25° and 12∠55°. Remember that the vectors must begin at the same point and that all angles are referenced off the x-axis (0°). To find the resultant of these two vectors, form a parallelogram using the vectors as two of the sides (see Figure A–14). The resultant is drawn from the corner of the parallelogram where the two vectors intersect to the opposite corner.

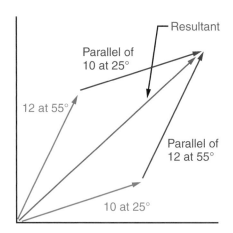

The parallelogram method.

A.12 Using the Rectangular Components

As was hinted at previously, vectors can be represented as the sum of their two components (called rectangular form or rectangular notation) or as a magnitude at an angle (polar notation). Figure A–15 shows two vectors.

Table A–3 shows the two ways that each vector can be represented. Be aware that the letter j (electricians use j instead of i) is itself a vector defined as $j = 1\angle 90°$.

TABLE A–3   Polar and Rectangular Formats for Vectors of Figure A–15

Vector	Polar	Rectangular
V_1	$6.4\angle 51.3°$	$4 + j5$
V_2	$3.6\angle 33.7°$	$3 + j2$

Example

Add the vectors V_1 and V_2 in Figure A–15.

Solution:
First the problem is solved by entering the polar form into a scientific calculator and adding the results (the Hewlett Packard model HP48GX was used for this example):

$$V_1 + V_2 = V_t = 6.4\angle 51.3° + 3.6\angle 33.7° = 9.89\angle 45°$$

Next, add the vectors using their rectangular components. The sum of the two vectors is equal to the sum of their horizontal components added to the sum of their vertical components:

$$V_1 + V_2 = V_t = (4 + j5) + (3 + j2) = (7 + j7)$$

Are the two answers the same? For the answer, refer to Figure A–16. This figure is the vector (V_T) as determined by adding the horizontal and vertical parts of the two component vectors. Notice that V_T forms a right triangle with the horizontal and vertical axes. This means that the tan of the unknown angle (??) is equal to 7/7. In other words, ?? = arctan(7/7) = 45°. What about the length? From the Pythagorean theorem, $V_t = \sqrt{7^2 + 7^2} = \sqrt{98} = 9.89$.

FIGURE A–15   Vector notation.

FIGURE A–16   Sum of vectors in Figure A–15.

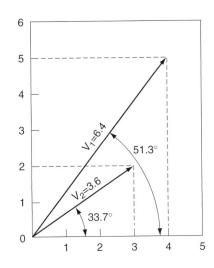

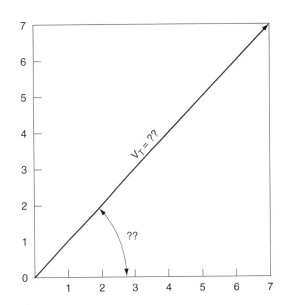

MULTIPLYING AND DIVIDING VECTORS

You will be called on to multiply and divide vectors in your work as an electrician. For example, the use of Ohm's law requires vectors for voltage, current, and impedance (AC resistance). To multiply two vectors, you multiply their magnitudes and add their angles. To divide two vectors, you divide their magnitudes and subtract their angles.

Example

Given: $V_1 = 5\angle 20°$ and $V_2 = 4\angle 30°$
Problem 1: What is the product of V_1 times V_2?

Solution:

$$V_1 \times V_2 = 5\angle 20 \times 4\angle 30 = (5 \times 4)\angle 20 + 30 = 20\angle 50$$

Problem 2: What is the quotient of V_1 divided by V_2?

Solution:

$$\frac{V_1}{V_2} = \frac{5\angle 20}{4\angle 30} = \left(\frac{5}{4}\right)\angle 20 - 30 = 1.25 - 10$$

SUMMARY

Scalar quantities have only magnitude. Vectors are lines that indicate both direction and magnitude.

Vectors can be added, subtracted, multiplied, or divided. The direction of a vector is indicated by the arrowhead at the end. When a single vector is produced from combining two or more vectors, it is called a *resultant*. From zero degrees, vectors rotate in a counterclockwise direction.

The sum of all the angles in a right triangle is 180°. The relationship between the length of the sides of a right triangle to the number of degrees in its angles can be expressed as the sine, cosine, or tangent of a particular angle.

The sine function is the relationship of the opposite side divided by the hypotenuse. The cosine function is the relationship of the adjacent side divided by the hypotenuse. The tangent function is the relationship of the opposite side divided by the adjacent side.

The hypotenuse is always the longest side of a right triangle. The opposite and adjacent sides are determined by which angle is being used as the reference angle.

AC—Alternating current A current that varies, or "alternates," from one polarity to another. Current flows from negative to positive in half of the cycle and reverses and flows from positive to negative in the other half of the cycle.

Ampere The measure of a specific number of electrons is called a coulomb. That number is approximately 6.25×10^{18}, or 6.25 billion electrons. When that number of electrons passes a specific point in 1 second, or 1 coulomb per second, we say that 1 amp is flowing and is represented by the abbreviation "A". In calculations, current is represented by the letter "I" representing the intensity of flow.

Armature The component in the process of generation that has voltage induced into it. It is not always the moving component but is the place where generated voltage is collected.

Armature reaction The bending of the field magnetic flux by a reaction to the armature magnetic field is known as armature reaction. Armature reaction causes a displacement of the neutral point with consequent arcing at the commutator and resultant voltage drop.

Apparent power The amount of power that is delivered to an AC circuit. It is calculated by the product of line voltage ($V_{Applied}$) multiplied by the line current (I_T). Apparent Power = VA = $V_{Applied} \times I_T$. Apparently it is the amount of power consumed, but is not the true power of a circuit with reactive components.

Average voltage The average voltage value of a AC since wave rectified to rippling DC. The value depends on if the AC is half-wave rectified or full-wave rectified. The value for half-wave rectified rippling DC is

$$V_{AVG\,(1/2\text{-wave})} = \frac{V_{Peak}}{\pi} = \frac{.637 \times V_{Peak}}{2}$$

The value for full-wave rectified rippling DC is

$$V_{AVG\,(full\text{-wave})} = \frac{2 \times V_{Peak}}{\pi} = 637 \times V_{Peak}$$

Brushes Sliding contacts usually made of a carbon or graphite alloy that are positioned so they are connected to the rotating armature winding segments. Often brushes are used with a commutator, therefore delivering the AC output created on the to the load as DC.

Capacitance 1. The ratio of charge to potential on an electrically charged, isolated conductor. 2. The unit of measure is the farad and the formula variable is "C". The ratio of the electric charge transferred from one to the other of a pair of conductors to the resulting potential difference between them. 3. The property of a circuit element that permits it to store charge. The part of the circuit exhibiting capacitance. (Excerpted from American Heritage Talking Dictionary. Copyright ©1997 The Learning Company, Inc. All Rights Reserved.)

Capacitor A capacitor is an electric circuit element used to store charge temporarily. In general it consists of two metallic plates separated and insulated from each other by a dielectric. A capacitor may also be referred to as a condenser. (Excerpted from American Heritage Talking Dictionary. Copyright ©1997 The Learning Company, Inc. All Rights Reserved.)

Capacitive reactance The opposition to AC current flow caused by a capacitor. The symbol for capacitive reactance is X_C. The formula to calculate capacitive reactance is

$$X_C = \frac{1}{2\,\pi fC}.$$

CEMF (Counter EMF) The voltage induced in an inductor by the changing magnetic field. The voltage induced is "counter to" the voltage, or in opposition to, the voltage that produced the original magnetic field.

Commutator A multi-segment rotating connection that is connected to the armature windings.

Compensating, interpole, or Thomson-Ryan windings Compensating windings, interpole windings, or Thomson-Ryan windings are field windings that are located physically between the main field poles. Interpoles are used to reduce armature reaction and are connected in series with armature current.

Conductor A material whose electrons can be moved with relative ease when voltage is applied.

Dielectric A nonconductor of electricity, especially a substance with electrical conductivity less than a millionth (10^{-6}) of a siemen. A dielectric has the ability to insulate against a current flow between conducting surfaces. (Excerpted from American Heritage Talking Dictionary. Copyright ©1997 The Learning Company, Inc. All Rights Reserved.)

Direct current A current that only flows in one direction without relationship to time.

Electromotive force The electrical pressure generated between two areas with different amounts of electrical charge. The unit of measure is the volt and represented in formulas with "E" representing electromotive force.

Electron flow theory The theory that electrons—electricity—flow from a negative potential toward a positive potential.

Excitation current The current supplied to the field of an AC generator. The excitation current creates the magnetic field that the armature cuts through.

Farad The unit of capacitance in the meter-kilogram-second (MKS) system equal to the capacitance of a capacitor having an equal and opposite charge of one coulomb on each plate and a potential difference of one volt between the plates. The larger the farad value, the more charge a capacitor can hold.

Flat compounding The effects of compounding in a DC generator so that the "no load" output voltage is the same as the "full load" voltage. The voltage vs. load curve on a graph is essentially flat, or no change in voltage, with change in load.

Field windings The part of the DC generator that creates the magnetic field that is cut by the armature windings.

Frequency The number of complete waveforms completed in 1 second. Frequency is a measure of how often the waveform is completed in relation to time, expressed in cycles per second. The unit of measure is the hertz (Hz) named after Heinrich Hertz.

Half-power point The point at which the load resistance is dissipating half the maximum power value that could occur at the load.

Henry Unit of measure for inductance. A coil has an inductance of 1 henry when a current change of 1 ampere per second causes an induced voltage of 1 volt. The henry is a value based on the physical characteristics of the inductor (coil). The unit of measure is the henry, abbreviated as H. The symbol for inductance used in formulas is L.

Hypotenuse The side of a right triangle that is opposite the right angle.

Impedance The vector sum of the oppositions found in some AC circuits. The vectors may include inductive reactance, capacitive reactance, and resistance. The vector addition of these components will result in the total opposition to the AC current. Impedance is measured in ohms and is represented by the letter Z in formulas.

Inductance The property of an electric circuit displayed when a varying current induces an electromotive force in that circuit or in a neighboring circuit. A circuit has inductance when magnetic induction is produced.

Inductive reactance The opposition to AC current flow caused by an inductor. The symbol for inductance is X_L. The formula to calculate inductive reactance is $X_L = 2\pi fL$.

Inductor Inductor is the name given to an electrical circuit component that exhibits the properties of inductance. If a coil of wire creates self-induction creating a counter electromotive force (CEMF), then it is referred to as an inductor.

Instantaneous current of a sine wave To find the current at any given moment of an AC sine wave use formula $I_{inst} = I_{Peak} \times \sin(\theta)$. For voltage use $E_{inst} = E_{Peak} \times \sin(\theta)$.

Insulator A material whose electrons strongly oppose movement from one atom to the next.

Isolation transformer An electrical transformer designed to have the same level of voltage at the secondary terminals that it has on the primary terminals. There is no electrical connection between the primary and secondary, creating a separate secondary circuit "isolated" from the original source.

Leakage flux Flux lines that do not link properly in an inductor. Some flux lines are lost to the surrounding space and do not complete the magnetic path. Leakage flux reduces the overall magnetic field and the resulting inductance.

Lenz's Law An electromagnetic field interacting with a conductor will generate electrical current that induces a counter magnetic field that opposes the magnetic field generating the current.

Magnet wire Magnet wire has a coating of insulating varnish that provides the voltage rating for the wire. The wire is specifically designed to be used in coil winding of motors, generators, or coils. The wire fits into slots for easy winding and is used to produce a magnetic field.

Magnetizing current The relatively small current that is used to magnetize the core of a transformer but does not provide for the secondary load current is considered magnetizing current or exciting current. It is the primary current when no electrical load is connected to the transformer secondary.

Microfarad 1×10^{-6} farads. Abbreviated as μF.

Micro-microfarad See picofarad. Abbreviated as μμF.

Nanofarad 1×10^{-9} farads. Abbreviated as nF. Not commonly used.

Ohms The unit of resistance in a circuit. Specifically, it is the amount of resistance that allows 1 ampere of current to flow when 1 volt is applied. The symbol used to represent the ohm is the Greek letter omega (Ω). In calculations, resistance is represented by the letter R. A component in a circuit that creates resistance is called a resistor.

Overcompounding A condition that occurs is a DC generator when the output voltage at "no load" is less than the output voltage at "full load". The output voltage increases as electrical load is added.

Peak-to-Peak The full value of the AC sine wave measured from the positive peak to the negative peak is referred to as the P-to-P value.

Periodic Table of the Elements A tabular arrangement of the elements according to their atomic numbers so that elements with similar properties are in the same column.

Permeability The ability of a material to concentrate or focus magnetic flux lines. Absolute permeability is measured in henries per meter. The permeability of a vacuum is $\frac{1.26 \times 10^{-6} \text{ H}}{\text{m}}$. The symbol used for magnetic permeability in a formula is the lowercase Greek letter mu (μ).

Phasor A vector that rotates. Phasors are used to describe voltages, currents, and other such quantities in electrical systems.

Picofarad 1×10^{-12} farads. Abbreviated as pF. Also called a micro-microfarad (abbreviated as μμF).

Piezoelectric Generation of electricity from pressure and pressure produced by electricity.

PMG Abbreviation for permanent magnet generator. In this type of generator, the field is provided by a very strong permanent magnet.

Polar form Representation of a vector as a magnitude and an angle.

Power factor A factor applied to the apparent power to yield true power. The power factor decimal is often expressed as a percentage. It is also the ratio of watts divided by volt-amps expressed as a percentage of true watts compared to apparent volt-amps.

Q of a coil The quality factor of a coil or the ratio of inductive reactance (X_L) to resistance (R). A formula to find Q is

$$Q = \frac{X_L}{R}.$$

Reactance The amount of opposition to current flow exhibited by a magnetic field in an inductor or an electrostatic field in a capacitor.

Reactive power The form of power that is produced by the reactive components of a circuit, such as the inductive or capacitive reactance. It is energy stored in a magnetic field or an electrostatic field and is returned to the circuit as the fields diminish. Even though this is a form of energy for the circuit no power is consumed. It is measured in volts-amps reactive (VARs) because it is caused by the reactive current, 90° out of phase with the voltage.

Rectangular form Representing a vector as the sum of two other vectors that are at right angles to each other.

Reluctance The opposition a material has to the flow of magnetic flux lines. If the material's magnetic domains are not easily aligned, the magnetic fields cannot easily pass through the material.

Resistance The amount of opposition to current flow through a material in a DC or AC circuit measured in ohms.

Resonance, or Resonant frequency The value in hertz at which inductive reactance and capacitive reactance are equal (in resonance) in a circuit that contains both capacitance and reactance. At series resonant frequency, the circuit impedance is at a minimum and the circuit current is at a maximum.

Revolving armature AC generator A generator that has the field windings on the stator and the armature windings on the rotor.

Revolving field generator A generator that has the field windings on the rotor and the armature windings on the stator.

RMS The RMS (root mean square) of an AC waveform is the square root of the mean, or average, of all the instantaneous values squared. The RMS gives us the effective value of the AC. This RMS value has the same effect as DC of the same value.

Round rotor A more streamlined type of construction where the coils are wound longitudinally on the rotor. This type of construction is generally much lighter and is used extensively on high-speed generators. Also called a turbo rotor.

Salient pole Salient means projecting outward; projecting or jutting beyond a line or surface; protruding. A salient pole rotor is a type of rotating field construction where the field poles are wound individually and mounted along the outside edge of the rotor.

Scalar A number with magnitude only. For example, 5 miles per hour is a scalar.

Self-excited generator Self-excited refers to the method in which the generator receives power to produce the magnetic field for generation. A portion of the output power is fed back to the generator's field as it provides its own excitation current.

Self-inductance The property of an electrical component (such as a coil of wire) to induce a voltage into itself as the current through the component changes.

Semiconductor A material with four electrons in the valence shell. A semiconductor has more electrical resistance than a conductor but less resistance than an insulator.

Skin effect In AC, the current that is forced to travel near the surface of a conductor. This is the results of eddy current repelling the current away from the center of the conductor. This effect creates the same consequence as reducing the cross-sectional area of a conductor. Both eddy current and skin effect is related to frequency and is a real factor at high frequencies—it is usually negligible at 60 Hz.

Slip Slip refers to the difference in speed measured in rpm between the synchronous speed of the rotating magnetic field of the stator, and the slower speed of the induction rotor. It is often expressed as a percent using the formula:

$$\% \text{ slip} = \frac{(\text{sync speed} - \text{actual speed})}{\text{sync speed} \times 100}$$

Slip rings Continuous bands of metal installed around a motor shaft. Slip rings are connected to the rotating windings and provide a path for the current to reach the brushes. Slip rings perform various functions such as providing a path for an excitation current to the rotor in a synchronous motors or removing current from AC generators.

Squirrel cage rotor A rotating member of an induction motor with an iron core and conductors around the periphery. The conductors have induced voltage and resultant current flow as the rotating magnetic field spins on the stator. The stripped down rotor, without iron and only the conducting bars, resembles a squirrel cage exercise wheel.

Stator The stationary part of a generator or motor. The stator consists of the core and the windings of copper wire. Cores are most often constructed using laminated electrical grade steel to reduce losses due to eddy currents. The purpose of the stator is to form a strong electromagnet in which the rotor turns. In some small motors, the windings are replaced with permanent magnets.

Step-down transformer An electrical transformer designed to have a lower voltage at the secondary terminals than it had at the primary terminals.

Step-up transformer An electrical transformer designed to have a higher voltage at the secondary output than it has at the primary input.

Subtractive or additive polarity Polarity refers to the secondary lead markings when compared to the primary lead markings. With a subtractive polarity transformer, if the leads are brought out so that when H1 and X1 are connected together, a voltmeter between H2 and X2 will read the difference between the primary and secondary. An additive polarity transformer indicates that if H1 and X2 are connected together a voltmeter between H2 and X1 will read the sum of the two coil voltages.

Synchronous speed The exact speed of a rotating member to produce a desired frequency for a generator. This same term is used to describe the speed that is produced by a specific frequency to make an induction motor run.

Synchroscope A type of instrument that is used to determine the phase relationship between two separate generator voltages. Often used for determining synchronization of two generators prior to connecting the outputs together.

True power The actual dissipated watts of a circuit. This is the form of energy expended by the conversion of electric energy into other forms of energy. It is the power measured in watts.

Undercompounding The condition that exists when the setup of the generator results in the "no load" voltage being greater than the "full load" voltage. As the generator supplies more load current, the output voltage decreases.

Unity power factor Unity power factor is a ratio of true power (watts) divided by apparent power (volt-amps) with the dividend as one, or the ratio is 1. Unity power factor is also known as 100%, meaning that 100% of the volt-amps are converted to watts and the reactive power does not affect the circuit. Power factor can not be over unity or 100%.

Valence ring The outermost ring, shell, or orbit of electrons in an atom.

Vector A number with both magnitude and direction. For example, 5 miles per hour going due east is a vector.

Volt The electromotive force that pushes electrons through the conductors, wires, or components of a circuit. It is similar to the pressure exerted on a system of fluid using pipes. The higher the pressure, the more flow. Specifically, the volt is the amount of work done per coulomb of charge (volts = joules per coulomb) and is represented by the letter V. In calculations, voltage is represented by the letter E. Remember that voltage is the force required in creating flow, but volts do not flow through the circuit.

Volts per turn Volts per turn is used as a reference point for the amount of magnetic flux produced by a coil. (The more volts per turn the more flux.) It is not a specific unit of measure but a way to relate the primary turns and voltage to the secondary turns and voltage.